T0280533

Vibrations and Waves

The Manchester Physics Series

General Editors
F.K. LOEBINGER: F. MANDL: D.J. SANDIFORD

School of Physics & Astronomy,
The University of Manchester

Properties of Matter:	B.H. Flowers and E. Mendoza
Statistical Physics: *Second Edition*	F. Mandl
Electromagnetism: *Second Edition*	I.S. Grant and W.R. Phillips
Statistics:	R.J. Barlow
Solid State Physics: *Second Edition*	J.R. Hook and H.E. Hall
Quantum Mechanics:	F. Mandl
Computing for Scientists:	R.J. Barlow and A.R. Barnett
The Physics of Stars: *Second Edition*	A.C. Phillips
Nuclear Physics	J.S. Lilley
Introduction to Quantum Mechanics	A.C. Phillips
Particle Physics: *Third Edition*	B.R. Martin and G. Shaw
Dynamics and Relativity	J.R. Forshaw and A.G. Smith
Vibrations and Waves	G.C. King

VIBRATIONS AND WAVES

George C. King
School of Physics & Astronomy,
The University of Manchester, Manchester, UK

A John Wiley and Sons, Ltd., Publication

This edition first published 2009

Registered office
John Wiley & Sons Ltd, The Atrium, Southern Gate, Chichester, West Sussex, PO19 8SQ, United Kingdom

For details of our global editorial offices, for customer services and for information about how to apply for permission to reuse the copyright material in this book please see our website at www.wiley.com.

Library of Congress Cataloging-in-Publication Data

King, George C.
Vibrations and waves / George C. King.
p. cm.
Includes bibliographical references and index.
ISBN 978-0-470-01188-1 – ISBN 978-0-470-01189-8
1. Wave mechanics. 2. Vibration. 3. Oscillations. I. Title.
QC174.22.K56 2009
531′.1133 – dc22

2009007660

A catalogue record for this book is available from the British Library

ISBN 978-0-470-01188-1 (HB)
ISBN 978-0-470-01189-8 (PB)

Typeset in 10/12 Times by Laserwords Private Limited, Chennai, India

Franz Mandl
(1923–2009)

This book is dedicated to Franz Mandl. I first encountered him as an inspirational teacher when I was an undergraduate. Later, we became colleagues and firm friends at Manchester. Franz was the editor throughout the writing of the book and made many valuable suggestions and comments based upon his wide-ranging knowledge and profound understanding of physics. Discussions with him about the various topics presented in the book were always illuminating and this interaction was one of the joys of writing the book.

Contents

Editors' Preface to the Manchester Physics Series

The Manchester Physics Series is a series of textbooks at first degree level. It grew out of our experience at the University of Manchester, widely shared elsewhere, that many textbooks contain much more material than can be accommodated in a typical undergraduate course; and that this material is only rarely so arranged as to allow the definition of a short self-contained course. In planning these books we have had two objectives. One was to produce short books so that lecturers would find them attractive for undergraduate courses, and so that students would not be frightened off by their encyclopaedic size or price. To achieve this, we have been very selective in the choice of topics, with the emphasis on the basic physics together with some instructive, stimulating and useful applications. Our second objective was to produce books which allow courses of different lengths and difficulty to be selected with emphasis on different applications. To achieve such flexibility we have encouraged authors to use flow diagrams showing the logical connections between different chapters and to put some topics in starred sections. These cover more advanced and alternative material which is not required for the understanding of latter parts of each volume.

Although these books were conceived as a series, each of them is self-contained and can be used independently of the others. Several of them are suitable for wider use in other sciences. Each Author's Preface gives details about the level, prerequisites, etc., of that volume.

The Manchester Physics Series has been very successful since its inception 40 years ago, with total sales of more than a quarter of a million copies. We are extremely grateful to the many students and colleagues, at Manchester and elsewhere, for helpful criticisms and stimulating comments. Our particular thanks go to the authors for all the work they have done, for the many new ideas they have contributed, and for discussing patiently, and often accepting, the suggestions of the editors.

Finally we would like to thank our publishers, John Wiley & Sons, Ltd, for their enthusiastic and continued commitment to the Manchester Physics Series.

F. K. Loebinger
F. Mandl
D. J. Sandiford
August 2008

Author's Preface

Vibrations and waves lie at the heart of many branches of the physical sciences and engineering. Consequently, their study is an essential part of the education of students in these disciplines. This book is based upon an introductory 24-lecture course on vibrations and waves given by the author at the University of Manchester. The course was attended by first-year undergraduate students taking physics or a joint honours degree course with physics. This book covers the topics given in the course although, in general, it amplifies to some extent the material delivered in the lectures.

The organisation of the book serves to provide a logical progression from the simple harmonic oscillator to waves in continuous media. The first three chapters deal with simple harmonic oscillations in various circumstances while the last four chapters deal with waves in their various forms. The connecting chapter (Chapter 4) deals with coupled oscillators which provide the bridge between waves and the simple harmonic oscillator. Chapter 1 describes simple harmonic motion in some detail. Here the universal importance of the simple harmonic oscillator is emphasised and it is shown how the elegant mathematical description of simple harmonic motion can be applied to a wide range of physical systems. Chapter 2 extends the study of simple harmonic motion to the case where damping forces are present as they invariably are in real physical situations. It also introduces the quality factor Q of an oscillating system. Chapter 3 describes forced oscillations, including the phenomenon of resonance where small forces can produce large oscillations and possibly catastrophic effects when a system is driven at its resonance frequency. Chapter 4 describes coupled oscillations and their representation in terms of the normal modes of the system. As noted above, coupled oscillators pave the way to the understanding of waves in continuous media. Chapter 5 deals with the physical characteristics of travelling waves and their mathematical description and introduces the fundamental wave equation. Chapter 6 deals with standing waves that are seen to be the normal modes of a vibrating system. A consideration of the general motion of a vibrating string as a superposition of normal modes leads to an introduction of the powerful technique of Fourier analysis. Chapter 7 deals with some of the most dramatic phenomena produced by waves, namely interference and diffraction. Finally, Chapter 8 describes the superposition of a group of waves to form a modulated wave or wave packet and the behaviour of this group of waves in a dispersive medium. Throughout the book, the fundamental principles of waves and vibrations are emphasised so that these principles can be applied to a wide range of oscillating systems and to a variety of waves including electromagnetic waves and sound waves. There are some topics that are not required for other parts of the book and these are indicated in the text.

Waves and vibrations are beautifully and concisely described in terms of the mathematical equations that are used throughout the book. However, emphasis is always placed on the physical meaning of these equations and undue mathematical complication and detail are avoided. An elementary knowledge of differentiation and integration is assumed. Simple differential equations are used and indeed waves and vibrations provide a particularly valuable way to explore the solutions of these differential equations and their relevance to real physical situations. Vibrations and waves are well described in complex representation. The relevant properties of complex numbers and their use in representing physical quantities are introduced in Chapter 3 where the power of the complex representation is also demonstrated.

Each chapter is accompanied by a set of problems that form an important part of the book. These have been designed to deepen the understanding of the reader and develop their skill and self-confidence in the application of the equations. Some solutions and hints to these problems are given at the end of the book. It is, of course, far more beneficial for the reader to try to solve the problems *before* consulting the solutions.

I am particularly indebted to Dr Franz Mandl who was my editor throughout the writing of the book. He read the manuscript with great care and physical insight and made numerous and valuable comments and suggestions. My discussions with him were always illuminating and rewarding and indeed interacting with him was one of the joys of writing the book. I am very grateful to Dr Michele Siggel-King, my wife, who produced all the figures in the book. She constructed many of the figures depicting oscillatory and wave motion using computer simulation programs and she turned my sketches into suitable figures for publication. I am also grateful to Michele for proofreading the manuscript. I am grateful to Professor Fred Loebinger who made valuable comments about the figures and to Dr Antonio Juarez Reyes for working through some of the problems.

George C. King

1

Simple Harmonic Motion

In the physical world there are many examples of things that vibrate or oscillate, i.e. perform periodic motion. Everyday examples are a swinging pendulum, a plucked guitar string and a car bouncing up and down on its springs. The most basic form of periodic motion is called simple harmonic motion (SHM). In this chapter we develop quantitative descriptions of SHM. We obtain equations for the ways in which the displacement, velocity and acceleration of a simple harmonic oscillator vary with time and the ways in which the kinetic and potential energies of the oscillator vary. To do this we discuss two particularly important examples of SHM: a mass oscillating at the end of a spring and a swinging pendulum. We then extend our discussion to electrical circuits and show that the equations that describe the movement of charge in an oscillating electrical circuit are identical in form to those that describe, for example, the motion of a mass on the end of a spring. Thus if we understand one type of harmonic oscillator then we can readily understand and analyse many other types. The universal importance of SHM is that to a good approximation many real oscillating systems behave like simple harmonic oscillators when they undergo oscillations of small amplitude. Consequently, the elegant mathematical description of the simple harmonic oscillator that we will develop can be applied to a wide range of physical systems.

1.1 PHYSICAL CHARACTERISTICS OF SIMPLE HARMONIC OSCILLATORS

Observing the motion of a pendulum can tell us a great deal about the general characteristics of SHM. We could make such a pendulum by suspending an apple from the end of a length of string. When we draw the apple away from its *equilibrium position* and release it we see that the apple swings back towards the equilibrium position. It starts off from rest but steadily picks up speed. We notice that it *overshoots* the equilibrium position and does not stop until it reaches the

Vibrations and Waves George C. King

other extreme of its motion. It then swings back toward the equilibrium position and eventually arrives back at its initial position. This pattern then repeats with the apple swinging backwards and forwards *periodically*. Gravity is the *restoring force* that attracts the apple back to its equilibrium position. It is the *inertia* of the mass that causes it to overshoot. The apple has kinetic energy because of its motion. We notice that its velocity is zero when its displacement from the equilibrium position is a maximum and so its kinetic energy is also zero at that point. The apple also has potential energy. When it moves away from the equilibrium position the apple's vertical height increases and it gains potential energy. When the apple passes through the equilibrium position its vertical displacement is zero and so all of its energy must be kinetic. Thus at the point of zero displacement the velocity has its maximum value. As the apple swings back and forth there is a continuous exchange between its potential and kinetic energies. These characteristics of the pendulum are common to all simple harmonic oscillators: (i) periodic motion; (ii) an equilibrium position; (iii) a restoring force that is directed towards this equilibrium position; (iv) inertia causing overshoot; and (v) a continuous flow of energy between potential and kinetic. Of course the oscillation of the apple steadily dies away due to the effects of dissipative forces such as air resistance, but we will delay the discussion of these effects until Chapter 2.

1.2 A MASS ON A SPRING

1.2.1 A mass on a horizontal spring

Our first example of a simple harmonic oscillator is a mass on a horizontal spring as shown in Figure 1.1. The mass is attached to one end of the spring while the other end is held fixed. The equilibrium position corresponds to the unstretched length of the spring and x is the displacement of the mass from the equilibrium position along the x-axis. We start with an idealised version of a real physical situation. It is idealised because the mass is assumed to move on a frictionless surface and the spring is assumed to be weightless. Furthermore because the motion is in the horizontal direction, no effects due to gravity are involved. In physics it is quite usual to start with a simplified version or model because real physical situations are normally complicated and hard to handle. The simplification makes the problem tractable so that an initial, idealised solution can be obtained. The complications, e.g. the effects of friction on the motion of the oscillator, are then added in turn and at each stage a modified and improved solution is obtained. This process invariably provides a great deal of physical understanding about the real system and about the relative importance of the added complications.

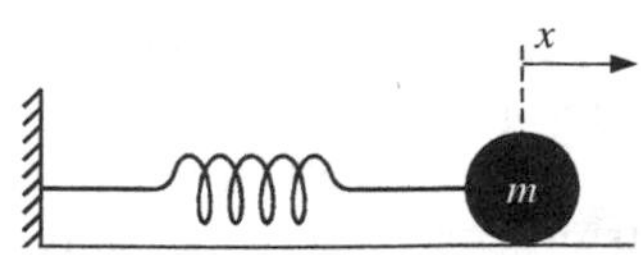

Figure 1.1 A simple harmonic oscillator consisting of a mass m on a horizontal spring.

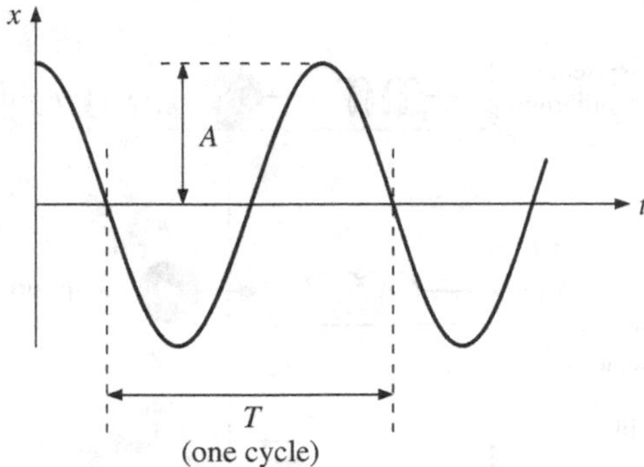

Figure 1.2 Variation of displacement x with time t for a mass undergoing SHM.

Experience tells us that if we pull the mass so as to extend the spring and then release it, the mass will move back and forth in a periodic way. If we plot the displacement x of the mass with respect to time t we obtain a curve like that shown in Figure 1.2. The *amplitude* of the oscillation is A, corresponding to the maximum excursion of the mass, and we note the *initial condition* that $x = A$ at time $t = 0$. The time for one complete cycle of oscillation is the period T. The frequency ν is the number of cycles of oscillation per unit time. The relationship between period and frequency is

$$\nu = \frac{1}{T}. \tag{1.1}$$

The units of frequency are hertz (Hz), where

$$1 \text{ Hz} \equiv 1 \text{ cycle per second} \equiv 1 \text{ s}^{-1}.$$

For small displacements the force produced by the spring is described by Hooke's law which says that the strength of the force is proportional to the extension (or compression) of the spring, i.e. $F \propto x$ where x is the displacement of the mass. The constant of proportionality is the spring constant k which is defined as the force per unit displacement. When the spring is extended, i.e. x is positive, the force acts in the opposite direction to x to pull the mass back to the equilibrium position. Similarly when the spring is compressed, i.e. x is negative, the force again acts in the opposite direction to x to push the mass back to the equilibrium position. This situation is illustrated in Figure 1.3 which shows the direction of the force at various points of the oscillation. We can therefore write

$$F = -kx \tag{1.2}$$

where the minus sign indicates that the force always acts in the opposite direction to the displacement. *All* simple harmonic oscillators have forces that act in this way: (i) the magnitude of the force is directly proportional to the displacement; and (ii) the force is always directed towards the equilibrium position.

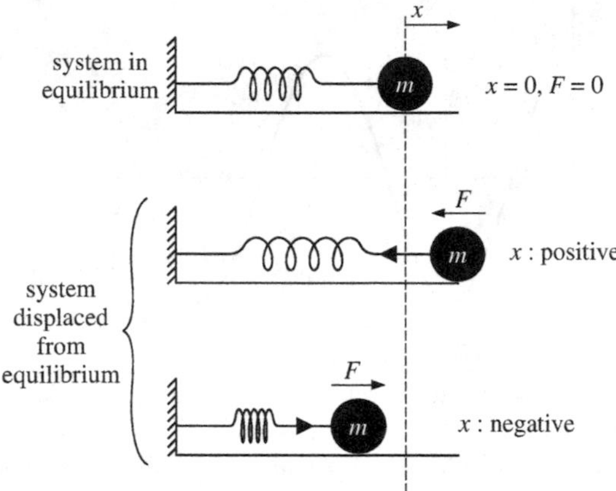

Figure 1.3 The direction of the force acting on the mass m at various values of displacement x.

The system must also obey Newton's second law of motion which states that the force is equal to mass m times acceleration a, i.e. $F = ma$. We thus obtain the equation of motion of the mass

$$F = ma = -kx. \tag{1.3}$$

Recalling that velocity v and acceleration a are, respectively, the first and second derivatives of displacement with respect to time, i.e.

$$a = \frac{\mathrm{d}v}{\mathrm{d}t} = \frac{\mathrm{d}^2x}{\mathrm{d}t^2}, \tag{1.4}$$

we can write Equation (1.3) in the form of the differential equation

$$m\frac{\mathrm{d}^2x}{\mathrm{d}t^2} = -kx \tag{1.5}$$

or

$$\boxed{\frac{\mathrm{d}^2x}{\mathrm{d}t^2} = -\omega^2 x} \tag{1.6}$$

where

$$\omega^2 = \frac{k}{m} \tag{1.7}$$

is a constant. Equation (1.6) is the equation of SHM and *all* simple harmonic oscillators have an equation of this form. It is a linear second-order differential equation; linear because each term is proportional to x or one of its derivatives and second order because the highest derivative occurring in it is second order. The reason for writing the constant as ω^2 will soon become apparent but we note that ω^2 is equal to the restoring force per unit displacement per unit mass.

1.2.2 A mass on a vertical spring

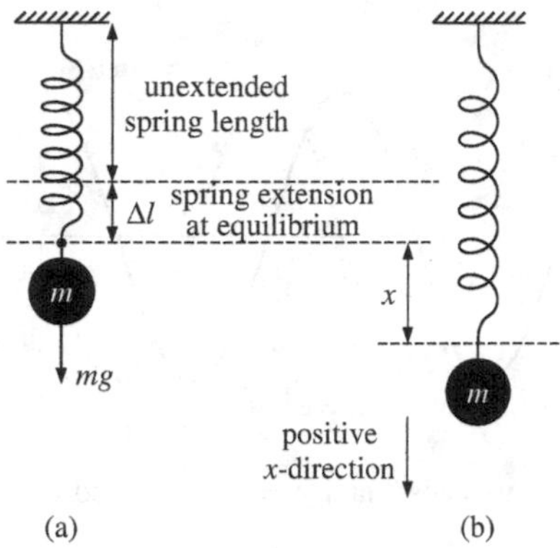

Figure 1.4 An oscillating mass on a vertical spring. (a) The mass at its equilibrium position. (b) The mass displaced by a distance x from its equilibrium position.

If we suspend a mass from a vertical spring, as shown in Figure 1.4, we have gravity also acting on the mass. When the mass is initially attached to the spring, the length of the spring increases by an amount Δl. Taking displacements in the downward direction as positive, the resultant force on the mass is equal to the gravitational force minus the force exerted upwards by the spring, i.e. the resultant force is given by $mg - k\Delta l$. The resultant force is equal to zero when the mass is at its equilibrium position. Hence

$$k\Delta l = mg.$$

When the mass is displaced downwards by an amount x, the resultant force is given by

$$F = m\frac{\mathrm{d}^2x}{\mathrm{d}t^2} = mg - k(\Delta l + x) = mg - k\Delta l - kx$$

i.e.

$$m\frac{\mathrm{d}^2x}{\mathrm{d}t^2} = -kx. \tag{1.8}$$

Perhaps not surprisingly, this result is identical to the equation of motion (1.5) of the horizontal spring: we simply need to measure displacements from the equilibrium position of the mass.

1.2.3 Displacement, velocity and acceleration in simple harmonic motion

To describe the harmonic oscillator, we need expressions for the displacement, velocity and acceleration as functions of time: $x(t)$, $v(t)$ and $a(t)$. These can be obtained by solving Equation (1.6) using standard mathematical methods. However, we will use our physical intuition to deduce them from the observed behaviour of a mass on a spring.

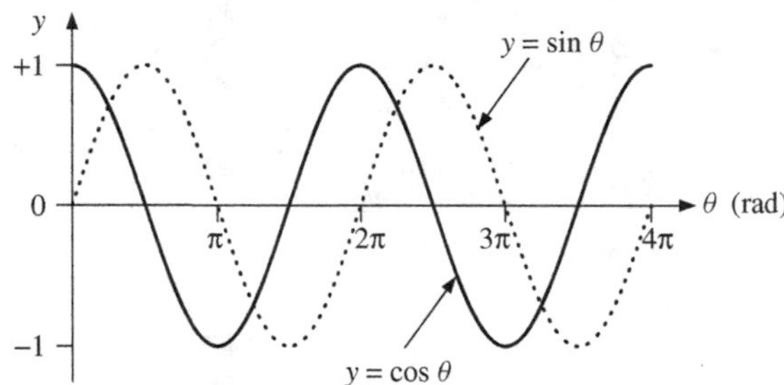

Figure 1.5 The functions $y = \cos\theta$ and $y = \sin\theta$ plotted over two complete cycles.

Observing the periodic motion shown in Figure 1.2, we look for a function $x(t)$ that also repeats periodically. Periodic functions that are familiar to us are $\sin\theta$ and $\cos\theta$. These are reproduced in Figure 1.5 over two complete cycles. Both functions repeat every time the angle θ changes by 2π. We can notice that the two functions are identical except for a shift of $\pi/2$ along the θ axis. We also note the initial condition that the displacement x of the mass equals A at $t = 0$. Comparison of the actual motion with the mathematical functions in Figure 1.5 suggests the choice of a cosine function for $x(t)$. We write it as

$$x = A\cos\left(\frac{2\pi t}{T}\right) \tag{1.9}$$

which has the correct form in that $(2\pi t/T)$ is an angle (in radians) that goes from 0 to 2π as t goes from 0 to T, and so repeats with the correct period. Moreover x equals A at $t = 0$ which matches the initial condition. We also require that $x = A\cos(2\pi t/T)$ is a solution to our differential equation (1.6). We define

$$\omega = \frac{2\pi}{T} \tag{1.10}$$

where ω is the *angular frequency* of the oscillator, with units of rad s^{-1}, to obtain

$$x = A\cos\omega t. \tag{1.11}$$

Then

$$\frac{dx}{dt} = v = -\omega A\sin\omega t, \tag{1.12}$$

and

$$\frac{d^2x}{dt^2} = a = -\omega^2 A\cos\omega t = -\omega^2 x. \tag{1.13}$$

So, the function $x = A\cos\omega t$ is a solution of Equation (1.6) and correctly describes the physical situation. The reason for writing the constant as ω^2 in Equation (1.6) is now apparent: the constant is equal to the square of the angular frequency of oscillation. We have also obtained expressions for the velocity v and acceleration a of the mass as functions of time. All three functions are plotted in Figure 1.6. Since they relate to different physical quantities, namely displacement, velocity and acceleration, they are plotted on separate sets of axes, although the time axes are aligned with respect to each other.

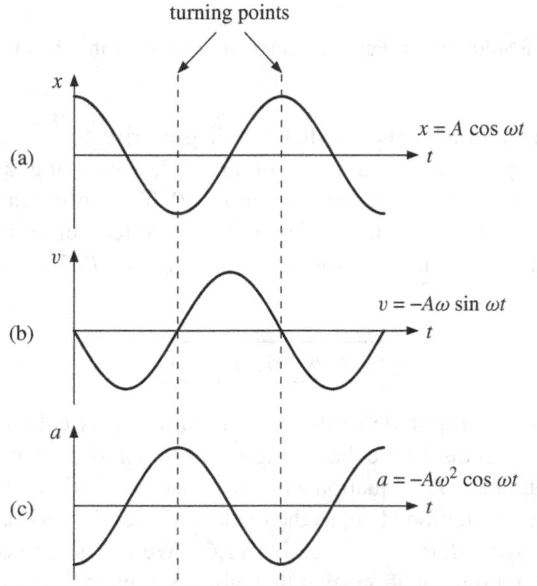

Figure 1.6 (a) The displacement x, (b) the velocity v and (c) the acceleration a of a mass undergoing SHM as a function of time t. The time axes of the three graphs are aligned.

Figure 1.6 shows that the behaviour of the three functions (1.11)–(1.13) agree with our observations. For example, when the displacement of the mass is greatest, which occurs at the *turning points* of the motion ($x = \pm A$), the velocity is zero. However, the velocity is at a maximum when the mass passes through its equilibrium position, i.e. $x = 0$. Looked at in a different way, we can see that the maximum in the velocity curve occurs before the maximum in the displacement curve by one quarter of a period which corresponds to an angle of $\pi/2$. We can understand at which points the maxima and minima of the acceleration occur by recalling that acceleration is directly proportional to the force. The force is maximum at the turning points of the motion but is of opposite sign to the displacement. The acceleration does indeed follow this same pattern, as is readily seen in Figure 1.6.

1.2.4 General solutions for simple harmonic motion and the phase angle ϕ

In the example above, we had the particular situation where the mass was released from rest with an initial displacement A, i.e. x equals A at $t = 0$. For the more

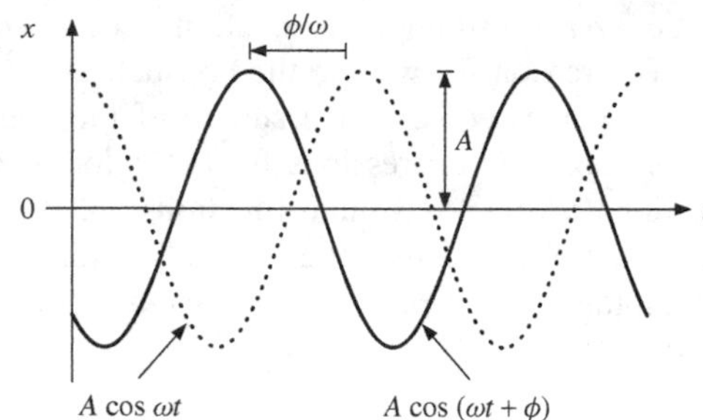

Figure 1.7 General solution for displacement x in SHM showing the phase angle ϕ, where $x = A\cos(\omega t + \phi)$.

general case, the motion of the oscillator will give rise to a displacement curve like that shown by the solid curve in Figure 1.7, where the displacement and velocity of the mass have arbitrary values at $t = 0$. This solid curve looks like the cosine function $x = A\cos\omega t$, that is shown by the dotted curve, but it is displaced horizontally to the left of it by a time interval $\phi/\omega = \phi T/2\pi$. The solid curve is described by

$$\boxed{x = A\cos(\omega t + \phi)} \tag{1.14}$$

where again A is the amplitude of the oscillation and ϕ is called the *phase angle* which has units of radians. [Note that changing ωt to $(\omega t - \phi)$ would shift the curve to the right in Figure 1.7.] Equation (1.14) is also a solution of the equation of motion of the mass, Equation (1.6), as the reader can readily verify. In fact Equation (1.14) is the *general solution* of Equation (1.6). We can state here a property of second-order differential equations that they always contain two arbitrary constants. In this case A and ϕ are the two constants which are determined from the initial conditions, i.e. from the position and velocity of the mass at time $t = 0$.

We can cast the general solution, Equation (1.14), in the alternative form

$$x = a\cos\omega t + b\sin\omega t, \tag{1.15}$$

where a and b are now the two constants. Equations (1.14) and (1.15) are entirely equivalent as we can show in the following way. Since

$$A\cos(\omega t + \phi) = A\cos\omega t\cos\phi - A\sin\omega t\sin\phi \tag{1.16}$$

and $\cos\phi$ and $\sin\phi$ have constant values, we can rewrite the right-hand side of this equation as

$$a\cos\omega t + b\sin\omega t,$$

where

$$a = A\cos\phi \text{ and } b = -A\sin\phi. \tag{1.17}$$

We see that if we add sine and cosine curves of the *same* angular frequency ω, we obtain another cosine (or corresponding sine curve) of angular frequency ω.

This is illustrated in Figure 1.8 where we plot $A\cos\omega t$ and $A\sin\omega t$, and also $(A\cos\omega t + A\sin\omega t)$ which is equal to $\sqrt{2}A\cos(\omega t - \pi/4)$. As the motion of a simple harmonic oscillator is described by sines and cosines it is called harmonic and because there is only a single frequency involved, it is called simple harmonic.

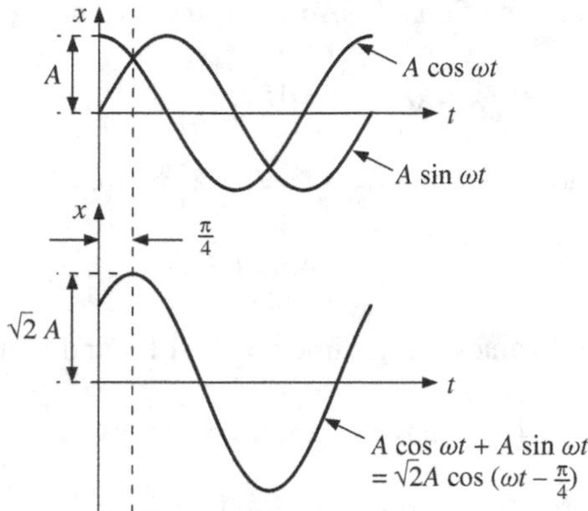

Figure 1.8 The addition of sine and cosine curves with the same angular frequency ω. The resultant curve also has angular frequency ω.

There is an important difference between the constants A and ϕ in the general solution for SHM given in Equation (1.14) and the angular frequency ω. The constants are determined by the initial conditions of the motion. However, the angular frequency of oscillation ω is determined only by the properties of the oscillator: the oscillator has a *natural frequency of oscillation* that is independent of the way in which we start the motion. This is reflected in the fact that the SHM equation, Equation (1.6), already contains ω which therefore has nothing to do with any particular solutions of the equation. This has important practical applications. It means, for example, that the period of a pendulum clock is independent of the amplitude of the pendulum so that it keeps time to a high degree of accuracy.[1] It means that the pitch of a note from a piano does not depend on how hard you strike the keys. For the example of the mass on a spring, $\omega = \sqrt{k/m}$. This expression tells us that the angular frequency becomes lower as the mass increases and becomes higher as the spring constant increases.

Worked example

In the example of a mass on a horizontal spring (cf. Figure 1.1) m has a value of 0.80 kg and the spring constant k is 180 N m^{-1}. At time $t = 0$ the mass is observed to be 0.04 m further from the wall than the equilibrium position and is moving away from the wall with a velocity of 0.50 m s^{-1}. Obtain an

[1] This assumes that the pendulum is operating as an ideal harmonic oscillator which is a good approximation for oscillations of small amplitude.

expression for the displacement of the mass in the form $x = A\ (\cos \omega t + \phi)$, obtaining numerical values for A, ω and ϕ.

Solution

The angular frequency ω depends only on the oscillator parameters k and m, and not on the initial conditions. Substituting their values gives

$$\omega = \sqrt{k/m} = 15.0 \text{ rad s}^{-1}$$

To find the amplitude A: From $x = A\cos(\omega t + \phi)$ we obtain

$$v = -A\omega \sin(\omega t + \phi).$$

Substituting the initial values (i.e. at time $t = 0$), of x and v into these equations gives

$$0.04 = A\cos\phi, \quad 0.50 = -15A\sin\phi.$$

From $\cos^2\phi + \sin^2\phi = 1$, we obtain $A = 0.052$ m.
To find the phase angle ϕ: Substituting the value for A leads to two equations for ϕ:

$$\cos\phi = 0.04/0.052, \qquad \text{giving } \phi = 39.8^\circ \text{ or } 320^\circ,$$

$$\sin\phi = -0.50/(15 \times 0.052), \qquad \text{giving } \phi = -39.8^\circ \text{ or } 320^\circ.$$

Since ϕ must satisfy both equations, it must have the value $\phi = 320^\circ$.
The angular frequency ω is given in rad s^{-1}. To convert ϕ to radians:

$$\phi = (\pi/180) \times 320 \text{ rad} = 5.59 \text{ rad. Hence, } x = 0.052\cos(15t + 5.59) \text{ m.}$$

1.2.5 The energy of a simple harmonic oscillator

Consideration of the energy of a system is a powerful tool in solving physical problems. For one thing, scalar rather than vector quantities are involved which usually simplifies the analysis. For the example of a mass on a spring, (Figure 1.1), the mass has kinetic energy K and potential energy U. The kinetic energy is due to the motion and is given by $K = \frac{1}{2}mv^2$. The potential energy U is the energy stored in the spring and is equal to the work done in extending or compressing it, i.e. 'force times distance'. The work done on the spring, extending it from x' to $x' + \mathrm{d}x'$, is $kx'\mathrm{d}x'$. Hence the work done extending it from its unstretched length by an amount x, i.e. its potential energy when extended by this amount, is

$$U = \int_0^x kx'\mathrm{d}x' = \frac{1}{2}kx^2. \tag{1.18}$$

Similarly, when the spring is compressed by an amount x the stored energy is again equal to $\frac{1}{2}kx^2$.

Conservation of energy for the harmonic oscillator follows from Newton's second law, Equation (1.5). In terms of the velocity v, this becomes

$$m\frac{\mathrm{d}v}{\mathrm{d}t} = -kx.$$

Multiplying this equation by $\mathrm{d}x = v\mathrm{d}t$ gives

$$mv\mathrm{d}v = -kx\mathrm{d}x$$

and since $\mathrm{d}(x^2) = 2x\mathrm{d}x$ and $\mathrm{d}(v^2) = 2v\mathrm{d}v$, we obtain

$$\mathrm{d}\left(\frac{1}{2}mv^2\right) = -\mathrm{d}\left(\frac{1}{2}kx^2\right).$$

Integrating this equation gives

$$\frac{1}{2}mv^2 + \frac{1}{2}kx^2 = \text{constant},$$

where the right-hand term is a constant of integration. The two terms on the left-hand side of this equation are just the kinetic energy K and the potential energy U of the oscillator. It follows that the constant on the right-hand side is the total energy E of the oscillator, i.e. we have derived conservation of energy for this case:

$$\boxed{E = K + U = \frac{1}{2}mv^2 + \frac{1}{2}kx^2} \tag{1.19}$$

Equation (1.19) enables us to calculate the energy E of the harmonic oscillator for any solution of the oscillator. If we take the general solution $x = A\cos(\omega t + \phi)$, we obtain the velocity

$$v = \frac{\mathrm{d}x}{\mathrm{d}t} = -\omega A\sin(\omega t + \phi) \tag{1.20}$$

and the potential and kinetic energies

$$U = \frac{1}{2}kx^2 = \frac{1}{2}kA^2\cos^2(\omega t + \phi) \tag{1.21}$$

$$K = \frac{1}{2}mv^2 = \frac{1}{2}m\omega^2A^2\sin^2(\omega t + \phi) = \frac{1}{2}kA^2\sin^2(\omega t + \phi) \tag{1.22}$$

where we substituted $\omega^2 = k/m$. Hence the total energy E is given by

$$\begin{aligned} E = K + U &= \frac{1}{2}kA^2[\sin^2(\omega t + \phi) + \cos^2(\omega t + \phi)] \\ &= \frac{1}{2}kA^2. \end{aligned} \tag{1.23}$$

Equation (1.23) shows that the energy of a harmonic oscillator is proportional to the square of the amplitude of the oscillation: the more we initially extend the spring the more potential energy we store in it. The first line of Equation (1.23) also shows that the energy of the system flows between kinetic and potential energies although the total energy remains constant. This is illustrated in Figure 1.9, which shows the variation of the potential and kinetic energies with time. We have taken $\phi = 0$ in this figure. We can also plot the kinetic and potential energies as functions of the displacement x. The potential energy $U = \frac{1}{2}kx^2$ is a parabola in x as shown in Figure 1.10. We do not need to work out the equivalent expression for the variation in kinetic energy since this must be equal to $(E - \frac{1}{2}kx^2)$ and is also shown in the figure.

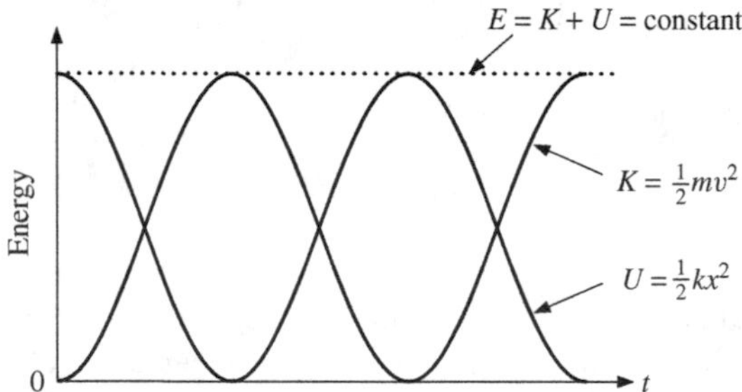

Figure 1.9 The variations of kinetic energy K and potential energy U with time t for a simple harmonic oscillator. The total energy of the oscillator E is the sum of the kinetic and potential energies and remains constant with time.

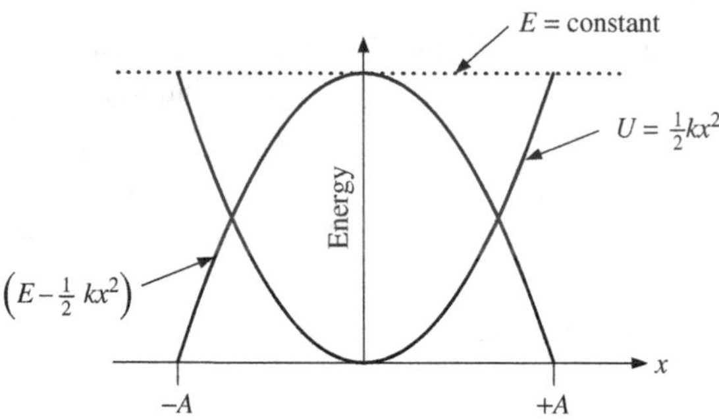

Figure 1.10 The variation of kinetic energy K and potential energy U with displacement x for a simple harmonic oscillator.

1.2.6 The physics of small vibrations

A mass on a spring is an example of a system in stable equilibrium. When the mass moves away from its equilibrium position the restoring force pulls or pushes it back. We found that the potential energy of a mass on a spring is proportional to x^2 so that the potential energy curve has the shape of a parabola given by

$U(x) = \frac{1}{2}kx^2$ (cf. Figure 1.10). This curve has a minimum when $x = 0$, which corresponds to the unstretched length of the spring. The movement of the mass is constrained by the spring and the mass is said to be confined in a potential well. The parabolic shape of this potential well gives rise to SHM. Any system that is in stable equilibrium will oscillate if it is displaced from its equilibrium state. We may think of a marble in a round-bottomed bowl. When the marble is pushed to one side it rolls back and forth in the bowl. The universal importance of the harmonic oscillator is that nearly all the potential wells we encounter in physical situations have a shape that is parabolic when we are sufficiently close to the equilibrium position. Thus, *most oscillating systems will oscillate with SHM when the amplitude of oscillation is small* as we shall prove in a moment. This situation is illustrated in Figure 1.11, which shows as a solid line the potential energy of a simple pendulum as a function of the angular displacement θ. (We will discuss the example of the simple pendulum in detail in Section 1.3.) Superimposed on it as a dotted line is a parabolic-shaped potential well, i.e. proportional to θ^2. Close to the equilibrium position ($\theta = 0$), the two curves lie on top of each other. So long as the amplitude of oscillation falls within the range where the two curves coincide the pendulum will execute SHM.

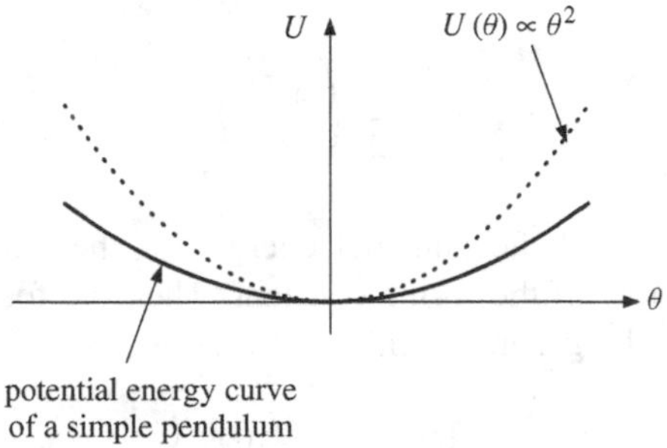

Figure 1.11 The solid curve represents the potential energy U of a simple pendulum as a function of its angular displacement θ. The dotted line represents the potential energy $U(\theta)$ of a simple harmonic oscillator for which the potential energy is proportional to θ^2. For small angular amplitudes, where the two curves overlap, a simple pendulum behaves as a simple harmonic oscillator.

We can see the above result mathematically using Taylor's theorem which says that any function $f(x)$ which is continuous and possesses derivatives of all orders at $x = a$ can be expanded in a power series in $(x - a)$ in the neighbourhood of the point $x = a$, i.e.

$$f(x) = f(a) + \frac{(x-a)}{1!}\left(\frac{\mathrm{d}f}{\mathrm{d}x}\right)_{x=a} + \frac{(x-a)^2}{2!}\left(\frac{\mathrm{d}^2 f}{\mathrm{d}x^2}\right)_{x=a} + \cdots \tag{1.24}$$

where the derivatives $\mathrm{d}f/\mathrm{d}x$, etc., are evaluated at $x = a$. (In practice all the potential wells that we encounter in physical situations can be described by functions that can be expanded in this way.) We see that Taylor's theorem gives the value of a function $f(x)$ in terms of the value of the function at $x = a$ and the values of the first and higher derivatives of x evaluated at $x = a$. If we expand $f(x)$ about $x = 0$, we have

$$f(x) = f(0) + x\left(\frac{\mathrm{d}f}{\mathrm{d}x}\right)_{x=0} + \frac{x^2}{2}\left(\frac{\mathrm{d}^2 f}{\mathrm{d}x^2}\right)_{x=0} + \cdots$$

In the case of a general potential well $U(x)$, we expand about the equilibrium position $x = 0$ to obtain

$$U(x) = U(0) + x\left(\frac{\mathrm{d}U}{\mathrm{d}x}\right)_{x=0} + \frac{x^2}{2}\left(\frac{\mathrm{d}^2 U}{\mathrm{d}x^2}\right)_{x=0} + \cdots \tag{1.25}$$

The first term $U(0)$ is a constant and has no physical significance in the sense that we can measure potential energy with respect to any position and indeed we can choose it to be equal to zero. The first derivative of U with respect to x is zero because the curve is a minimum at $x = 0$. The second derivative of U with respect to x, evaluated at $x = 0$, will be a constant. Thus if we retain only the first non-zero term in the expansion, which is a good approximation so long as x is small, we have

$$U(x) = \frac{x^2}{2}\left(\frac{\mathrm{d}^2 U}{\mathrm{d}x^2}\right)_{x=0} \tag{1.26}$$

This is indeed the form of the potential energy for the mass on a spring with $\mathrm{d}^2U/\mathrm{d}x^2$ playing the role of the spring constant. Then the force close to the equilibrium position takes the general form

$$F = -\frac{\mathrm{d}U}{\mathrm{d}x} = -x\left(\frac{\mathrm{d}^2 U}{\mathrm{d}x^2}\right)_{x=0} \tag{1.27}$$

The force is directly proportional to x and acts in the opposite direction which is our familiar result for the simple harmonic oscillator.

The fact that a vibrating system will behave like a simple harmonic oscillator when its amplitude of vibration is small means that our physical world is filled with examples of SHM. To illustrate this diversity Table 1.1 gives examples of a variety of physical systems that can oscillate and their associated periods of oscillation. These examples occur in both classical and quantum mechanics. Clearly the more massive the system, the greater is the period of oscillation. For the case of a vibrating tuning fork, we can tell that the ends of the fork are oscillating at a single frequency because we hear a pure note that we can use to tune musical instruments. A plucked guitar string will also oscillate and indeed musical instruments provide a wealth of examples of SHM. These oscillations, however, will in general be more complicated than that of the tuning fork but even here these complex oscillations are a superposition of SHMs as we shall see in Chapter 6. The balance wheel of a mechanical clock, the sloshing of water in a lake and the swaying of a sky scraper in the wind provide further examples of classical oscillators.

TABLE 1.1 Examples of systems that can oscillate and the associated periods of oscillation.

System	Period (s)
Sloshing of water in a lake	$\sim 10^2 - 10^4$
Large bridges and buildings	$\sim 1 - 10$
A clock pendulum or balance wheel	~ 1
String instruments	$\sim 10^{-3} - 10^{-2}$
Piezoelectric crystals	$\sim 10^{-6}$
Molecular vibrations	$\sim 10^{-15}$

A good example of SHM in the microscopic world is provided by the vibrations of the atoms in a crystal. The forces between the atoms result in the regular lattice structure of the crystal. Furthermore, when an atom is slightly displaced from its equilibrium position it is subject to a net restoring force. The shape of the resultant potential well approximates to a parabola for small amplitudes of vibration. Thus when the atoms vibrate they do so with SHM. Einstein used a simple harmonic oscillator model of a crystal to explain the observed variation of heat capacity with temperature (see also Mandl,[2] Section 6.2). He assumed that the atoms were harmonic oscillators that vibrate independently of each other but with the same angular frequency and he used a quantum mechanical description of these oscillators. As we have seen, in classical mechanics the energy of an oscillator is proportional to the square of the amplitude and can take any value, i.e. the energy is continuous. A fundamental result of quantum mechanics is that the energy of a harmonic oscillator is quantised, i.e. only a discrete set of energies is possible. Einstein's quantum model predicted that the specific heat of a crystal, such as diamond, goes to zero as the temperature of the crystal decreases, unlike the classical result that the specific heat is independent of temperature. Experiment shows that the specific heat of diamond does indeed go to zero at low temperatures.

Another example of SHM in quantum physics is provided by the vibrations of the two nuclei of a hydrogen molecule. The solid curve in Figure 1.12 represents the potential energy U of the hydrogen molecule as a function of the separation r between the nuclei, where we have taken the potential energy to be zero at infinite separation. This potential energy is due to the Coulomb interaction of the electrons and nuclei and the quantum behaviour of the electrons. The curve exhibits a minimum at $r_0 = 0.74 \times 10^{-10}$ m. At small separation ($r \to 0$) the potential energy tends to infinity, representing the strong repulsion between the nuclei. The nuclei perform oscillations about the equilibrium separation. The dotted line in Figure 1.12 shows the parabolic form of the potential energy of a harmonic oscillator, centred at the equilibrium seperation r_0. For small amplitudes of oscillation (i.e. when the nuclei are not too highly excited) the vibrations occur within the range where the two curves coincide. Again, according to quantum mechanics, only a discrete set of vibrational energies is possible. For a simple harmonic oscillator with angular frequency ω the only allowed values of the energy are $\frac{1}{2}\hbar\omega$, $\frac{3}{2}\hbar\omega$, $\frac{5}{2}\hbar\omega$, ..., where

[2] Statistical Physics, F. Mandl, Second Edition, 1988, John Wiley & Sons, Ltd.

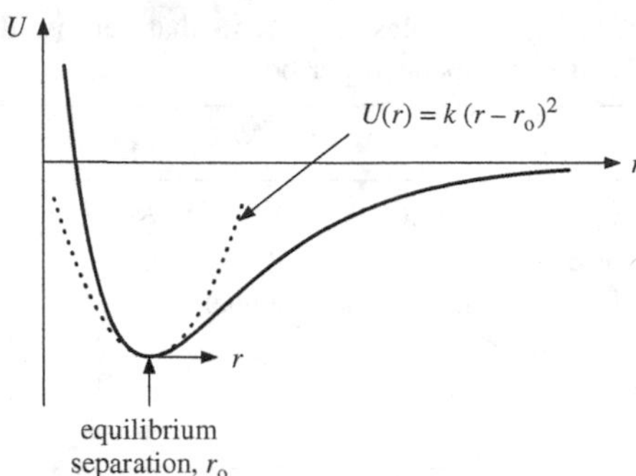

Figure 1.12 The solid curve represents the variation of potential energy of a hydrogen molecule as a function of the separation of the two hydrogen nuclei. The dotted curve represents the potential energy of a simple harmonic oscillator centred on the equilibrium separation r_o of the two nuclei.

$\hbar$ is Planck's constant divided by 2π. The observed vibrational line spectra of molecules correspond to transitions between these energy levels with the emission of electromagnetic radiation that typically lies in the infrared part of the electromagnetic spectrum. These spectra provide valuable information about the properties of the molecule such as the strength of the molecular bond.

Worked example

The H_2 molecule has a vibrational frequency ν of 1.32×10^{14} Hz. Calculate the strength of the molecular bond, i.e. the 'spring constant', assuming that the molecule can be modelled as a simple harmonic oscillator.

Solution

In previous cases, we considered a mass vibrating at one end of a spring while the other end of the spring was connected to a rigid wall. Now we have two nuclei vibrating against each other, which we model as two equal masses connected by a spring. We can solve this new situation by realising that there is no translation of the molecule during the vibration, i.e. the centre of mass of the molecule does not move. Thus as one hydrogen nucleus moves in one direction by a distance x the other must move in the opposite direction by the same amount and of course both vibrate at the same frequency. The total extension is $2x$ and the tension in the 'spring' is equal to $2kx$ where k represents the 'spring constant' or bond strength. The equation of motion of each nucleus of mass m is then given by

$$m\frac{d^2x}{dt^2} = -2kx$$

or

$$\frac{m}{2}\frac{d^2x}{dt^2} = -kx. \tag{1.28}$$

This equation is analogous to Equation (1.5) where m has been replaced by $m/2$ which is called the *reduced mass* of the system. The classical angular frequency of vibration ω of the molecule is then equal to $\sqrt{2k/m}$. The frequency of vibration $\nu = 1/T = \omega/2\pi$ and $m = 1.67 \times 10^{-27}$ kg. Therefore

$$k = 4\pi^2\nu^2\frac{m}{2} = \frac{4\pi^2(1.32 \times 10^{14})^2 1.67 \times 10^{-27}}{2} = 574 \text{ N m}^{-1}.$$

1.3 THE PENDULUM

1.3.1 The simple pendulum

Timing the oscillations of a pendulum has been used for centuries to measure time accurately. The simple pendulum is the idealised form that consists of a point mass m suspended from a massless rigid rod of length l, as illustrated in Figure 1.13. For an angular displacement θ, the displacement of the mass *along* the arc of the circle of length l is $l\theta$. Hence the angular velocity along the arc is $l\mathrm{d}\theta/\mathrm{d}t$ and the angular acceleration is $l\mathrm{d}^2\theta/\mathrm{d}t^2$. At a displacement θ there is a tangential force on the mass acting along the arc that is equal to $-mg\sin\theta$, where as usual the minus sign indicates that it is a restoring force. Hence by Newton's second law we obtain

$$\frac{\mathrm{d}^2\theta}{\mathrm{d}t^2} = -\frac{g}{l}\sin\theta. \tag{1.29}$$

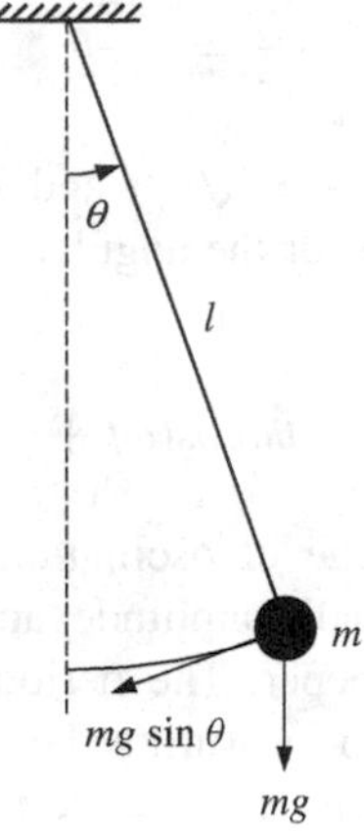

Figure 1.13 The simple pendulum of mass m and length l.

This equation does not have the same form as the equation of SHM, Equation (1.6), as we have $\sin\theta$ on the right-hand side instead of θ. However we can expand $\sin\theta$

in a power series in θ:

$$\sin\theta = \theta - \frac{\theta^3}{3!} + \frac{\theta^5}{5!} + \cdots. \tag{1.30}$$

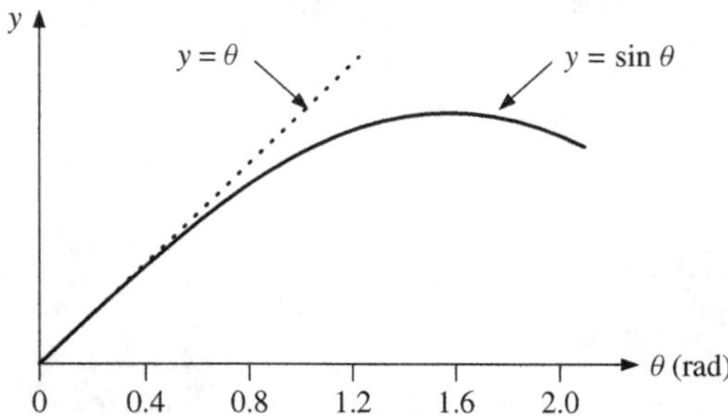

Figure 1.14 A comparison of the functions $y = \theta$ and $y = \sin\theta$ plotted against θ.

For small angular deflections the second and higher terms are much smaller than the first term. For example, if θ is equal to 0.1 rad (5.7°), which is typical for a pendulum clock, then the second term is only 0.17% of the first term and the higher terms are much smaller still. We can see this directly by plotting the functions $y = \sin\theta$ and $y = \theta$ on the same set of axes, as shown in Figure 1.14. The two curves are indistinguishable for values of θ below about $\frac{1}{4}$ rad (~15°). Thus for small values of θ, we need retain only the first term in the expansion (1.30) and replace $\sin\theta$ with θ (in radians) to give

$$\frac{d^2\theta}{dt^2} = -\frac{g}{l}\theta. \tag{1.31}$$

This is the equation of SHM with $\omega = \sqrt{g/l}$ and $T = 2\pi\sqrt{l/g}$, and we can immediately write down an expression for the angular displacement θ of the pendulum:

$$\theta = \theta_0 \cos(\omega t + \phi) \tag{1.32}$$

where θ_0 is the angular amplitude of oscillation. The period is independent of amplitude for oscillations of small amplitude and this is why the pendulum is so useful as an accurate time keeper. The period does, however, depend on the acceleration due to gravity and so measuring the period of a pendulum provides a way of determining the value of g. (In practice real pendulums do not have their mass concentrated at a point as in the simple pendulum as will be described in Section 1.3.3. So for an accurate determination of g a more sophisticated pendulum has been developed called the *compound pendulum*.) We finally note that for $l =$ 1.00 m and for a value of $g = 9.87$ m s^{-2}, the period of a simple pendulum is equal to $2\pi\sqrt{1.00/9.87} = 2.00$ s and indeed the second was originally defined as equal to one half the period of a 1 m simple pendulum.

1.3.2 The energy of a simple pendulum

We can also analyse the motion of the simple pendulum by considering its kinetic and potential energies. The geometry of the simple pendulum is shown in Figure 1.15. (The horizontal distance $x = l\sin\theta$ is not exactly the same as the distance along the arc, which is equal to $l\theta$. However, since $\sin\theta \simeq \theta$ for small θ, the difference is negligible.) From the geometry we have

$$l^2 = (l - y)^2 + x^2 \tag{1.33}$$

which gives

$$2ly = y^2 + x^2. \tag{1.34}$$

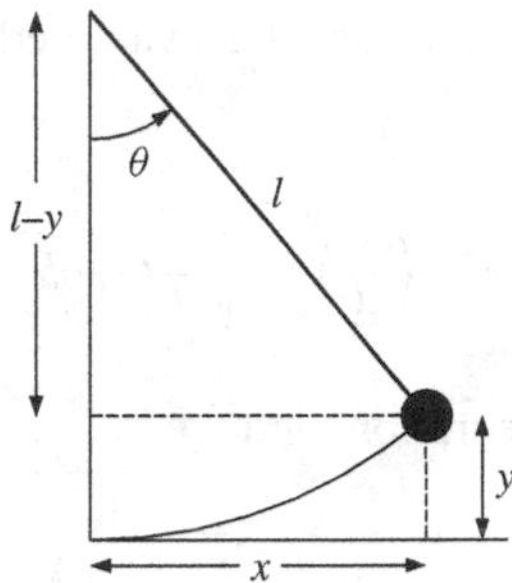

Figure 1.15 The geometry of the simple pendulum.

For small displacements of the pendulum, i.e. $x \ll l$, it follows that $y \ll x$, so that the term y^2 can be neglected and we can write,

$$y = \frac{x^2}{2l}. \tag{1.35}$$

As the mass is displaced from its equilibrium position its vertical height increases and it gains potential energy. This gain in potential energy is equal to $mgy = mgx^2/2l$. The total energy of the system E is given by the sum of the kinetic and potential energies:

$$E = K + U = \frac{1}{2}mv^2 + \frac{1}{2}\frac{mgx^2}{l}. \tag{1.36}$$

At the turning point of the motion, when x equals A, the velocity v is zero giving

$$E = \frac{1}{2}\frac{mgA^2}{l}. \tag{1.37}$$

From conservation of energy, it follows that

$$\frac{mgA^2}{l} = mv^2 + \frac{mgx^2}{l} \tag{1.38}$$

is true for all times. We can use Equation (1.38) to obtain expressions for velocity v and displacement x:

$$v = \frac{\mathrm{d}x}{\mathrm{d}t} = \sqrt{\frac{g(A^2 - x^2)}{l}}. \tag{1.39}$$

This expression describes how the velocity changes with the displacement x in SHM in contrast to Equation (1.12) which describes how the velocity changes with time t. Since $v = \mathrm{d}x/\mathrm{d}t$ we have

$$\int \frac{\mathrm{d}x}{\sqrt{A^2 - x^2}} = \sqrt{\frac{g}{l}} \int \mathrm{d}t. \tag{1.40}$$

The integral on the left-hand side can be evaluated using the substitution $x = A \sin\theta$, giving

$$\sin^{-1}\left(\frac{x}{A}\right) = \sqrt{\frac{g}{l}}t + \phi, \tag{1.41}$$

where ϕ is the constant of integration, and

$$x = A \sin\left(\sqrt{\frac{g}{l}}t + \phi\right). \tag{1.42}$$

Equation (1.42) describes SHM with $\omega = \sqrt{g/l}$ and $T = 2\pi\sqrt{l/g}$ as before.

At this point we note the similarity in the expressions for the total energy of the two examples of SHM that we have considered.

For the mass on a spring: $$E = \frac{1}{2}mv^2 + \frac{1}{2}kx^2. \tag{1.43a}$$

For the simple pendulum: $$E = \frac{1}{2}mv^2 + \frac{1}{2}\frac{mg}{l}x^2. \tag{1.43b}$$

Both expressions have the form: $$E = \frac{1}{2}\alpha v^2 + \frac{1}{2}\beta x^2, \tag{1.43c}$$

where α and β are constants. It is a universal characteristic of simple harmonic oscillators that their total energy can be written as the sum of two parts, one involving the (velocity)^2 and the other the (displacement)^2. Just as $m\mathrm{d}^2x/\mathrm{d}t^2 = -kx$, Equation (1.5), is the signature of SHM in terms of forces, Equation (1.43) is the signature of SHM in terms of energies. If we obtain either of these equations in the analysis of a system then we know we have SHM. We stress that the equations are the same for all simple harmonic oscillators: only the labels for the physical quantities change. We do not need to repeat the analysis again: we can simply take over the results already obtained. The constant α corresponds to the inertia of the system through which it can store kinetic energy. The constant β corresponds to the restoring force per unit displacement through which the system can store potential energy. When we differentiate the conservation of energy equation for SHM, Equation (1.43c), with respect to time we obtain

$$\frac{dE}{dt} = \alpha v \frac{dv}{dt} + \beta x \frac{dx}{dt} = 0$$

giving

$$\frac{d^2x}{dt^2} = -\frac{\beta}{\alpha}x.$$

Comparing this with Equation (1.6), it follows that the angular frequency of oscillation ω is equal to $\sqrt{\beta/\alpha}$.

Worked example

A marble of radius r rolls back and forth without slipping in a spherical dish of radius R. Use energy considerations to show that the motion is simple harmonic for small displacements of the marble from its equilibrium position and deduce an expression for the period of the oscillations. The moment of inertia I of a solid sphere of mass m about an axis through its centre is equal to $\frac{2}{5}mr^2$.

Solution

The equilibrium and displaced positions of the marble are shown in Figure 1.16, where the arrows indicate the rotation of the marble when it rotates through an angle ϕ. If the marble were rotating through an angle ϕ on a *flat* surface it would roll a distance $r\phi$. However on a spherical surface as in Figure 1.16, it rolls a distance l along the arc of radius R given by $l = r(\phi + \theta)$. Since $l = R\theta$,

$$\phi = \frac{(R-r)}{r}\theta \text{ and } \frac{d\phi}{dt} = \frac{(R-r)}{r}\left(\frac{d\theta}{dt}\right).$$

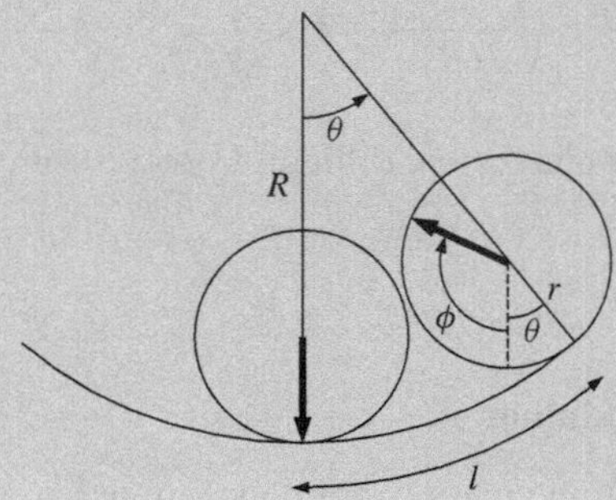

Figure 1.16 A marble of radius r that rolls back and forth without slipping in a spherical dish of radius R.

The total kinetic energy of the marble, as it moves along the surface of the dish, is equal to the kinetic energy of the translational motion of its centre of mass plus the kinetic energy of its rotational motion about the centre of mass. Hence

$$K = \frac{1}{2}mv^2 + \frac{1}{2}I\left(\frac{d\phi}{dt}\right)^2.$$

The translational kinetic energy is given by

$$\frac{1}{2}mv^2 = \frac{1}{2}m(R-r)^2\left(\frac{d\theta}{dt}\right)^2.$$

Therefore,

$$K = \frac{1}{2}m\left(\frac{7}{5}\right)(R-r)^2\left(\frac{d\theta}{dt}\right)^2$$

where we have substituted for I. The potential energy is

$$U = mg(R-r)(1-\cos\theta) = \frac{1}{2}mg(R-r)\theta^2$$

for small θ. Thus

$$E = \frac{1}{2}m\left(\frac{7}{5}\right)(R-r)^2\left(\frac{d\theta}{dt}\right)^2 + \frac{1}{2}mg(R-r)\theta^2.$$

This has the general form of the energy equation (1.43c) of a harmonic oscillator

$$E = \frac{1}{2}\alpha\left(\frac{d\theta}{dt}\right)^2 + \frac{1}{2}\beta\theta^2$$

where now θ represents the displacement coordinate. Hence

$$\omega = \sqrt{\frac{\beta}{\alpha}} = \sqrt{\frac{5g}{7(R-r)}} \text{ and } T = 2\pi\sqrt{\frac{7(R-r)}{5g}}.$$

This example would be much more difficult to solve from force considerations.

1.3.3 The physical pendulum

In a physical pendulum the mass is not concentrated at a point as in the simple pendulum, but is distributed over the whole body. It is thus more representative of real pendulums. An example of a physical pendulum is shown in Figure 1.17. It consists of a uniform rod of length l that pivots about a horizontal axis at its upper end. This is a rotating system where the pendulum rotates about its point of suspension. For a rotating system, Newton's second law for linear systems,

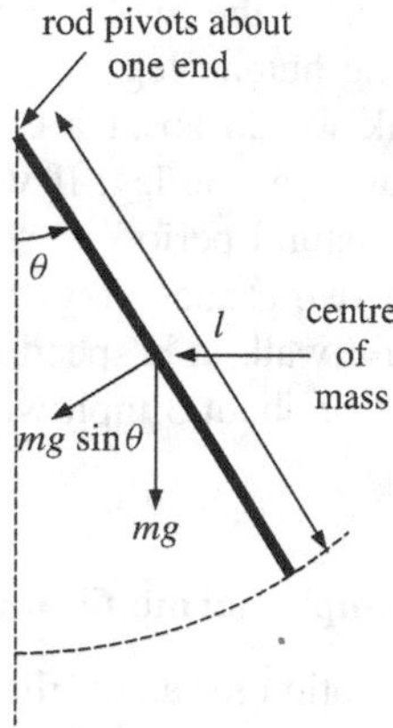

Figure 1.17 A rod that pivots about one of its ends, which is an example of a physical pendulum.

$m\mathrm{d}^2x/\mathrm{d}t^2 = F$, becomes

$$I\frac{\mathrm{d}^2\theta}{\mathrm{d}t^2} = \tau \tag{1.44}$$

where I is the moment of inertia of the body about its axis of rotation and τ is the applied torque. The moment of inertia of a uniform rod of length l about an end is equal to $\frac{1}{3}ml^2$ and its centre of mass is located at its mid point. The resultant torque τ on the rod when it is displaced through an angle θ is given by the product of the torque arm $\frac{1}{2}l$ and the component of the force normal to the torque arm ($mg\sin\theta$), i.e.

$$\tau = \left(\frac{1}{2}l\right) \times (-mg\sin\theta).$$

Hence we obtain

$$\frac{1}{3}ml^2\frac{\mathrm{d}^2\theta}{\mathrm{d}t^2} = -\frac{1}{2}mgl\sin\theta \tag{1.45}$$

giving

$$\frac{\mathrm{d}^2\theta}{\mathrm{d}t^2} = -\frac{3g}{2l}\sin\theta. \tag{1.46}$$

Again we can use the small-angle approximation to obtain

$$\frac{\mathrm{d}^2\theta}{\mathrm{d}t^2} = -\frac{3g}{2l}\theta. \tag{1.47}$$

This is SHM with $\omega = \sqrt{3g/2l}$ and $T = 2\pi\sqrt{2l/3g}$.

In a simple model we can describe the walking pace of a person in terms of a physical pendulum. We model the human leg as a solid rod that pivots from the hip. Furthermore, when we walk we do so at a comfortable pace that coincides with the natural period of oscillation of the leg. If we assume a value of 0.8 m for l, the length of the leg, then its natural period is $\sim$1.5 s. One complete period of the swinging leg corresponds to two strides. Try this yourself. If the length of a stride is, say, 1 m then we would walk at a speed of approximately 2/1.5 m s^{-1} which corresponds to 4.8 km h^{-1} or about 3 mph which is in good agreement with reality.

1.3.4 Numerical solution of simple harmonic motion[3]

When solving the equation of motion for an oscillating pendulum we made use of the small-angle approximation, $\sin\theta \simeq \theta$ when θ is small. This made the equation of motion much easier to solve. However an alternative way, without resorting to the small-angle approximation, is to solve the equation numerically. The essential idea is that if we know the position and velocity of the mass at time t and we know the force acting on it then we can use this knowledge to obtain good estimates of these parameters at time $(t + \delta t)$. We then repeat this process, step by step, over the full period of the oscillation to trace out the displacement of the mass with time. We can make these calculations as accurate as we like by making the time interval δt sufficiently small. To demonstrate this approach we apply it to the simple pendulum. Figure 1.18 shows a simple pendulum and the angular position of the mass at three instants of time each separated by δt, i.e. at t, $(t + \delta t)$ and $(t + 2\delta t)$. Using the notation $\dot{\theta}(t)$ and $\ddot{\theta}(t)$ for $\mathrm{d}\theta(t)/\mathrm{d}t$ and $\mathrm{d}^2\theta(t)/\mathrm{d}t^2$, respectively, we can write the equation of motion of the mass, Equation (1.29),

$$\ddot{\theta}(t) = -\frac{g}{l}\sin\theta(t). \tag{1.48}$$

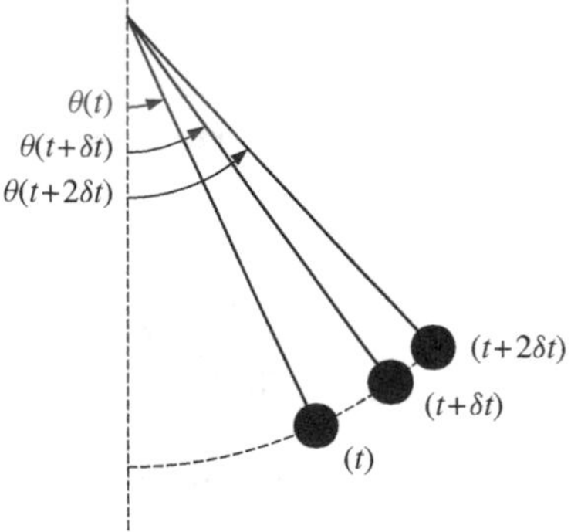

Figure 1.18 A simple pendulum showing the position of the mass at three instants of time separated by time interval δt.

[3] This section may be omitted as it is not required later in the book.

If the angular position of the mass is $\theta(t)$ at time t, then its position at time $(t + \delta t)$ will be different by an amount equal to the angular velocity of the mass times the time interval δt (cf. the familiar expression $x = vt$ for linear motion). We might be tempted to use $\dot{\theta}(t)$ for this angular velocity. However, as we know, the angular velocity varies during the time δt. A better estimate for the angular velocity is its *average* value between the times t and $(t + \delta t)$, i.e. $\dot{\theta}(t + \delta t/2)$. Thus to a good approximation we have

$$\theta(t + \delta t) = \theta(t) + \delta t \times \dot{\theta}(t + \delta t/2). \tag{1.49}$$

In a similar way we can relate the angular velocities of the mass at times separated by time δt, i.e. the new velocity will be different from the old value by an amount equal to $\delta t \times \ddot{\theta}(t)$, where $\ddot{\theta}(t)$ is the angular acceleration (cf. the familiar expression $v = u + at$ for linear motion). The acceleration also varies with time and so again we will use its average value during the time interval δt. For the evaluation of $\dot{\theta}(t + \delta t/2)$ this translates to

$$\dot{\theta}(t + \delta t/2) = \dot{\theta}(t - \delta t/2) + \delta t \times \ddot{\theta}(t) \tag{1.50}$$

where $\ddot{\theta}(t)$ is the average value of the angular acceleration between the times $(t - \delta t/2)$ and $(t + \delta t/2)$ which we know from Equation (1.48). For the first step of this calculation we need the value of the angular velocity at time $t = \delta t/2$. For this particular case we use the expression

$$\dot{\theta}(\delta t/2) = (\delta t/2) \times \ddot{\theta}(0). \tag{1.51}$$

Armed with these expressions for angular position, velocity and acceleration we can trace the angular displacement of the mass step by step.

We proceed by building up a table of consecutive values of $\theta(t), \dot{\theta}(t)$ and $\ddot{\theta}(t)$. As an example we chose the length of the simple pendulum to give $T = 2.0$ s and $\omega = \pi$. We also chose a time interval δt of 0.02 s (equal to one hundredth of the period) and an angular amplitude θ_0 of 0.10 rad (5.7°). The values obtained in the first 10 steps of the calculation are shown in Table 1.2 and were obtained using a hand calculator. For comparison the final column of Table 1.2 shows the values obtained from the analytic solution $\theta(t) = \theta_0 \cos \omega t$. We see that the numerically calculated values of the displacement are in agreement with the analytic values up to the third significant figure. These two sets of values for a complete period of oscillation are plotted in Figure 1.19 and show the familiar variation of displacement with time. The solid curve corresponds to the values of displacement obtained from the analytic solution $\theta(t) = \theta_0 \cos \omega t$, while the dots (•) correspond to the numerically computed values. The agreement is so good that the dots lie exactly on top of the analytic curve. These results demonstrate that the small angle approximation is valid in this case and that the numerical method gives accurate results.

This numerical method allows us to explore what happens for large-amplitude oscillations where the small angle approximation is no longer valid. Figure 1.20 shows the results for a very large angular amplitude of 1.0 rad (57°) which were

TABLE 1.2 Computed values of angular displacement, velocity and acceleration of a simple pendulum. The last column on the right shows the values obtained from the analytic solution.

Time (s)	Angular displacement, $\theta(t)$ (rad)	Angular acceleration, $\ddot{\theta}(t)$ (rad s^{-2})	Angular velocity, $\dot{\theta}(t)$ (rad s^{-1})	$\theta(t) = 0.1\cos\pi t$ (rad)
0.00	0.1000	−0.985	−0.0099	0.1000
0.02	0.0998	−0.983	−0.0295	0.0998
0.04	0.0992	−0.978	−0.0491	0.0992
0.06	0.0982	−0.968	−0.0685	0.0982
0.08	0.0968	−0.954	−0.0876	0.0969
0.10	0.0950	−0.937	−0.106	0.0951
0.12	0.0929	−0.915	−0.124	0.0930
0.14	0.0904	−0.891	−0.142	0.0905
0.16	0.0876	−0.863	−0.159	0.0876

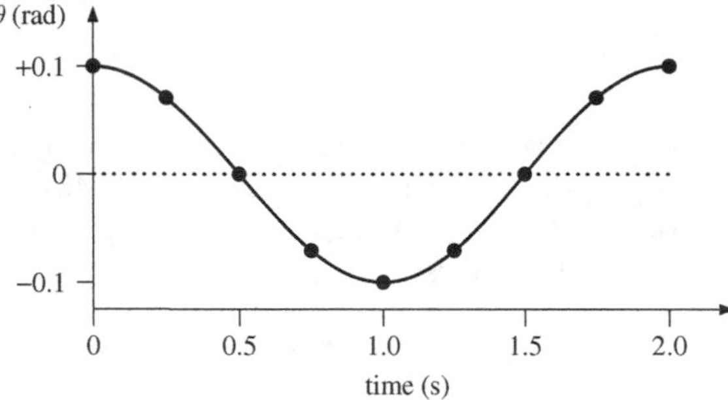

Figure 1.19 The angular displacement θ, plotted against time, for a simple pendulum with a small amplitude of oscillation; $\theta_0 = 0.1$ rad. The solid curve corresponds to the values of displacement obtained from the analytic solution $\theta(t) = \theta_0 \cos \omega t$, while the dots (•) correspond to the numerically computed values. The agreement is so good that the computed values lie on top of the analytical curve.

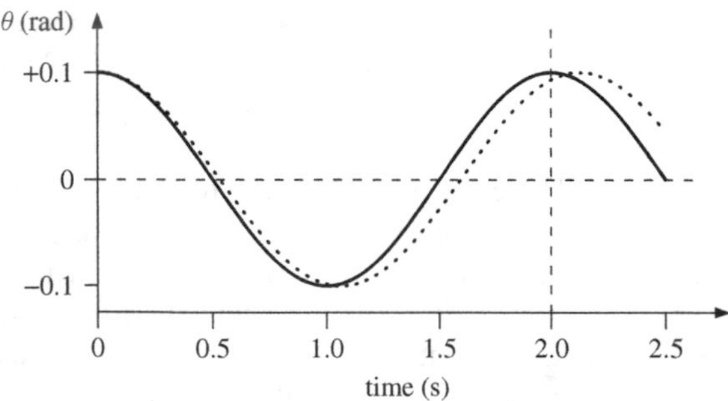

Figure 1.20 The angular displacement θ, plotted against time, of a simple pendulum for a large amplitude of oscillation; $\theta_0 = 1.0$ rad. The solid curve corresponds to the values of displacement obtained from the solution $\theta(t) = \theta_0 \cos \omega t$, while the dotted curve is obtained from the numerically computed results. For large-amplitude oscillations the period of the pendulum is no longer independent of amplitude and increases with amplitude.

obtained using a spreadsheet program. The solid curve corresponds to the values of displacement obtained from the solution $\theta(t) = \theta_0 \cos \omega t$ while the dotted curve is the one obtained from the numerically computed values. There is a significant difference between the two curves: the actual angular displacement of the mass, which is given by the numerical values, no longer closely matches the analytic solution. In particular the time period for the actual oscillations has increased to a value of 2.13 s: an increase of 6.5%. We see that for large-amplitude oscillations the period of the pendulum is no longer independent of amplitude and that it increases with amplitude.

1.4 OSCILLATIONS IN ELECTRICAL CIRCUITS: SIMILARITIES IN PHYSICS

In this section we consider oscillations in an electrical circuit. What we find is that these oscillations are described by a differential equation that is identical in form to Equation (1.6) and so has an identical solution: only the physical quantities associated with the differential equation are different. This illustrates that when we understand one physical situation we can understand many others. It also means that we can simulate one system by another and in this way build analogue computers, i.e. we can build an electrical circuit consisting of resistors, capacitors and inductors that will exactly simulate the operation of a mechanical system.

1.4.1 The *LC* circuit

The simplest example of an oscillating electrical circuit consists of an inductor L and capacitor C connected together in series with a switch as shown in Figure 1.21.

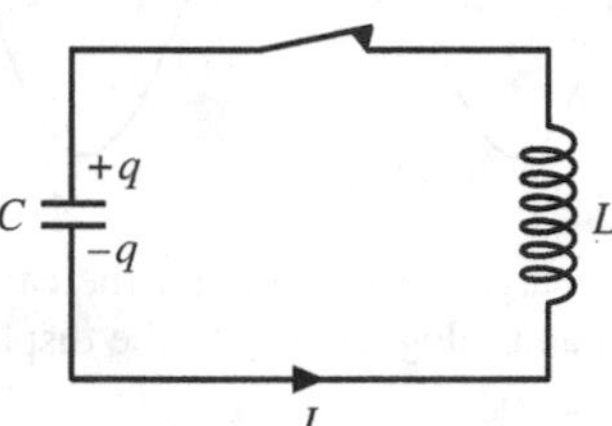

Figure 1.21 An electrical oscillator consisting of an inductor L and a capacitor C connected in series.

As usual we start with an idealised situation where we assume that the resistance in the circuit is negligible. This is analogous to the assumption for mechanical systems that there are no frictional forces present. Initially, the switch is open and the capacitor is charged to voltage V_C. The charge q on the capacitor is given by $q = V_C C$ where C is the capacitance. When the switch is closed the charge begins to flow through the inductor and a current $I = \mathrm{d}q/\mathrm{d}t$ flows in the circuit. This is a time-varying current and produces a voltage across the inductor given by $V_L = L\mathrm{d}I/\mathrm{d}t$. We can analyse the LC circuit using *Kirchhoff's law*, which states that 'the sum of the voltages around the circuit is zero', i.e. $V_C + V_L = 0$. Therefore

$$\frac{q}{C} + L\frac{\mathrm{d}I}{\mathrm{d}t} = 0 \tag{1.52}$$

giving

$$\frac{q}{C} + L\frac{\mathrm{d}^2q}{\mathrm{d}t^2} = 0 \tag{1.53}$$

and

$$\frac{\mathrm{d}^2q}{\mathrm{d}t^2} = -\frac{1}{LC}q. \tag{1.54}$$

This equation describes how the charge on a plate of the capacitor varies with time. It is of the same form as Equation (1.6) and represents SHM. The frequency of the oscillation is given directly by, $\omega = \sqrt{1/LC}$. Since we have the initial condition that the charge on the capacitor has its maximum value at $t = 0$, then the solution to Equation (1.54) is $q = q_0 \cos \omega t$, where q_0 is the initial charge on the capacitor. The variation of charge q with respect to t is shown in Figure 1.22 and is analogous to the way the displacement of a mass on a spring varies with time.

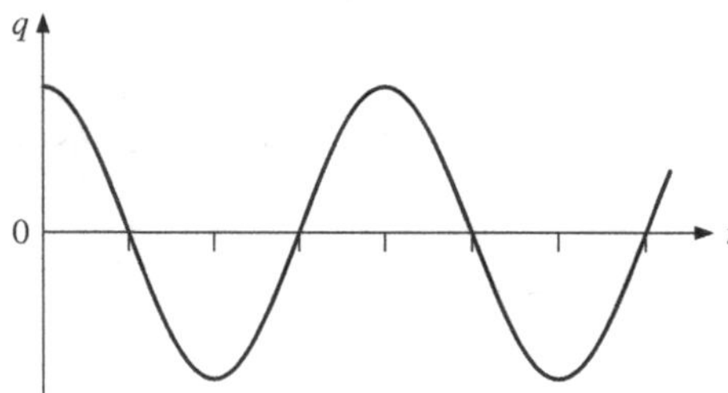

Figure 1.22 The variation of charge q with time on the capacitor in a series LC circuit. The charge oscillates in time in an analogous way to the displacement of a mass oscillating at the end of a spring.

We can also consider the energy of this electrical oscillator. The energy stored in a capacitor charged to voltage V_C is equal to $\frac{1}{2}CV_C^2$. This is electrostatic energy. The energy stored in an inductor is equal to $\frac{1}{2}LI^2$ and this is magnetic energy. Thus the total energy in the circuit is given by

$$E = \frac{1}{2}LI^2 + \frac{1}{2}CV_C^2 \tag{1.55}$$

or

$$E = \frac{1}{2}LI^2 + \frac{1}{2}\frac{q^2}{C}. \tag{1.56}$$

For these electrical oscillations the charge flows between the plates of the capacitor and through the inductor, so that there is a continuous exchange between electrostatic and magnetic energy.

1.4.2 Similarities in physics

We note the similarities between the equations for the mechanical and electrical cases

$$m\frac{\mathrm{d}^2x}{\mathrm{d}t^2} = -kx, \quad L\frac{\mathrm{d}^2q}{\mathrm{d}t^2} = -\frac{1}{C}q \tag{1.57a}$$

and

$$E = \frac{1}{2}m\left(\frac{\mathrm{d}x}{\mathrm{d}t}\right)^2 + \frac{1}{2}kx^2, \qquad E = \frac{1}{2}L\left(\frac{\mathrm{d}q}{\mathrm{d}t}\right)^2 + \frac{1}{2}\frac{q^2}{C}, \tag{1.57b}$$

where we have written $\mathrm{d}x/\mathrm{d}t$ for the velocity v and $\mathrm{d}q/\mathrm{d}t$ for the current I, in order to bring out more sharply the similarity of the two cases. In both cases we have the identical forms

$$\alpha\frac{\mathrm{d}^2Z}{\mathrm{d}t^2} = -\beta Z, \qquad E = \frac{1}{2}\alpha\left(\frac{\mathrm{d}Z}{\mathrm{d}t}\right)^2 + \frac{1}{2}\beta Z^2, \tag{1.58}$$

where α and β are constants and $Z = Z(t)$ is the oscillating quantity (see also Equations 1.43). In the mechanical case Z stands for the displacement x, and in the electrical case for the charge q. Thus all we have learned about mechanical oscillators can be carried over to electrical oscillators. Moreover we can see a direct correspondence between the two sets of physical quantities involved:

- q takes the place of x;
- L takes the place of m;
- $1/C$ takes the place of k.

For example, the inductance L is the electrical analogue of mechanical inertia m. These analogies enable us to build an electrical circuit that exactly mimics the operation of a mechanical system. This is useful because in the development of a mechanical system it is much easier to change, for example, the value of a capacitor in the analogue circuit than to manufacture a new mechanical component.

PROBLEMS 1

1.1 A mass of 0.50 kg hangs from a light spring and executes SHM so that its position x is given by $x = A\cos\omega t$. It is found that the mass completes 20 cycles of oscillation in 80 s. (a) Determine (i) the period of the oscillations, (ii) the angular frequency of the oscillations and (iii) the spring constant k. (b) Using a value of $A = 2$ mm, make sketches of the variations with time t of the displacement, velocity and acceleration of the mass.

1.2 The ends of a tuning fork oscillate at a frequency of 440 Hz with an amplitude of 0.50 mm. Determine (a) the maximum velocity and (b) the maximum acceleration of the ends.

1.3 A platform oscillates in the vertical direction with SHM. Its amplitude of oscillation is 0.20 m. What is the maximum frequency (Hz) of oscillation for a mass placed on the platform to remain in contact with the platform? (Assume $g = 9.81$ m s^{-2}.)

1.4 A mass executes SHM at the end of a light spring. (a) What fraction of the total energy of the system is potential and what fraction is kinetic at the instant when the displacement of the mass is equal to half the amplitude? (b) If the maximum amplitude of the oscillations is doubled, what will be the change in (i) the total energy of the system, (ii) the maximum velocity of the mass and (iii) the maximum acceleration of the mass. Will the period of oscillation change?

1.5 A mass of 0.75 kg is attached to one end of a horizontal spring of spring constant 400 N m^{-1}. The other end of the spring is attached to a rigid wall. The mass is pushed so that at time $t = 0$ it is 4.0 cm closer to the wall than the equilibrium position and is travelling towards the wall with a velocity of 0.50 m s^{-1}. (a) Determine the total energy of the oscillating system. (b) Obtain an expression for the displacement of the mass in the form $x = A\cos(\omega t + \phi)$ m, giving numerical values for A, ω and ϕ.

1.6

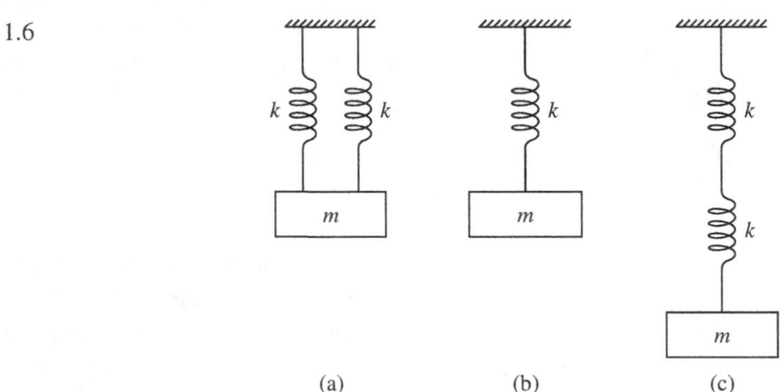

The figure shows three systems of a mass m suspended by light springs that all have the same spring constant k. Show that the frequencies of oscillation for the three systems are in the ratio $\omega_a : \omega_b : \omega_c = \sqrt{2} : 1 : \sqrt{1/2}$.

1.7 A test tube is weighted by some lead shot and floats upright in a liquid of density ρ. When slightly displaced from its equilibrium position and released, the test tube oscillates with SHM. (a) Show that the angular frequency of the oscillations is equal to $\sqrt{A\rho g/m}$ where g is the acceleration due to gravity, A is the cross-sectional area of the test tube and m is its mass. (b) Show that the potential energy of the system is equal to $\frac{1}{2}A\rho g x^2$ where x is the displacement from equilibrium. Hence give an expression for the total energy of the oscillating system in terms of the instantaneous displacement and velocity of the test tube.

1.8 We might assume that the period of a simple pendulum depends on the mass m, the length l of the string and g the acceleration due to gravity, i.e. $T \propto m^\alpha l^\beta g^\gamma$, where α, β and γ are constants. Consider the dimensions of the quantities involved to deduce the values of α, β and γ and hence show $T \propto \sqrt{l/g}$.

1.9 A simple pendulum has a length of 0.75 m. The pendulum mass is displaced horizontally from its equilibrium position by a distance of 5.0 mm and then released. Calculate (a) the maximum speed of the mass and (b) the time it takes to reach this speed. (Assume $g = 9.81$ m s^{-2}.)

1.10

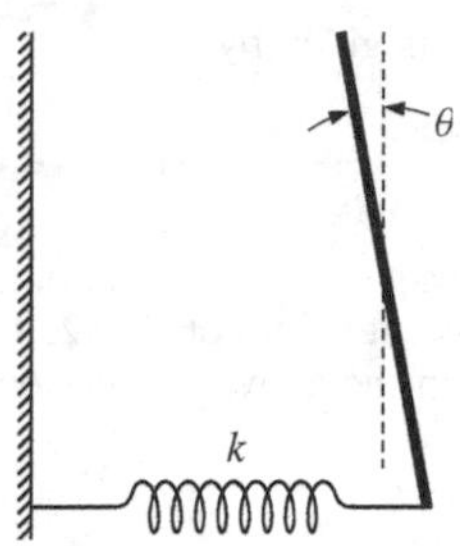

The figure shows a thin uniform rod of mass M and length $2L$ that is pivoted without friction about an axis through its mid point. A horizontal light spring of spring constant k is attached to the lower end of the rod. The spring is at its equilibrium length when the angle θ with respect to the vertical is zero. Show that for oscillations of small amplitude, the rod will undergo SHM with a period of $2\pi\sqrt{M/3k}$. The moment of inertia of the rod about its mid point is $ML^2/3$. (Assume the small angle approximations: $\sin\theta \simeq \theta$ and $\cos\theta \simeq 1$.)

1.11 The potential energy $U(x)$ between two atoms in a diatomic molecule can be expressed approximately by

$$U(x) = -\frac{a}{x^6} + \frac{b}{x^{12}}$$

where x is the separation of the atoms and a and b are constants. (a) Obtain an expression for the force between the two atoms and hence show that the equilibrium separation x_o of the atoms is equal to $(2b/a)^{1/6}$. (b) Show that the system will oscillate with SHM when slightly displaced from equilibrium with a frequency equal to $\sqrt{k/m}$, where m is the reduced mass and $k = 36a(a/2b)^{4/3}$.

1.12 A mass M oscillates at the end of a spring that has spring constant k and finite mass m. (a) Show that the total energy E of the system for oscillations of small amplitude is given by

$$E = \frac{1}{2}(M + m/3)v^2 + \frac{1}{2}kx^2$$

where v and x are the velocity and displacement of the mass M, respectively. (Hint: To find the kinetic energy of the spring, consider it to be divided into infinitesimal elements of length dl and find the total kinetic energy of these elements, assuming that the mass of the spring is evenly distributed along its length. The total energy E of the system is the sum of the kinetic energies of the spring and the mass M and the potential energy of the extended spring.) (b) Hence show that the frequency of the oscillations is equal to $\sqrt{k/(M + m/3)}$.

1.13 A particle oscillates with amplitude A in a one-dimensional potential $U(x)$ that is symmetric about $x = 0$, i.e. $U(x) = U(-x)$. (a) Show, from energy considerations, that the velocity v of the particle at displacement x from the equilibrium position ($x = 0$), is given by

$$v = \sqrt{2[U(A) - U(x)]/m}.$$

(b) Hence show that the period of oscillation T is given by

$$T = 4\sqrt{\frac{m}{2U(A)}}\int_0^A \frac{\mathrm{d}x}{\sqrt{[1 - U(x)/U(A)]}}.$$

(c) If the potential $U(x)$ is given by

$$U(x) = \alpha x^n$$

where α is a constant and $n = 2, 4, 6, \ldots$, obtain the dependence of the period T on the amplitude A for different values of $n = 2, 4, \ldots$. (Hint: Introduce the new variable of integration $\xi = x/A$ in the above expression for the period T.)

2

The Damped Harmonic Oscillator

In our description of an apple swinging back and forth at the end of a string (Section 1.1) we noted that this oscillating system is not ideal. After we set the apple in motion, the amplitude of oscillation steadily reduces and the apple eventually comes to rest. This is because there are dissipative forces acting and the system steadily loses energy. For example, the apple will experience a frictional force as it moves through the air. The motion is damped and such damped oscillations are the subject of this chapter. All real oscillating systems are subject to damping forces and will cease to oscillate if energy is not fed back into them. Often these damping forces are linearly proportional to velocity. Fortunately, this linear dependence leads to an equation of motion that can be readily solved to obtain solutions that describe the motion for various degrees of damping. Clearly the rate at which the oscillator loses energy will depend on the degree of damping and this is described by the *quality* of the oscillator. At first sight, damping in an oscillator may be thought undesirable. However, there are many examples where a controlled amount of damping is used to quench unwanted oscillations. Damping is added to the suspension system of a car to stop it bouncing up and down long after it has passed over a bump in the road. Additional damping was installed on London's Millennium Bridge shortly after it opened because it suffered from undesirable oscillations.

2.1 PHYSICAL CHARACTERISTICS OF THE DAMPED HARMONIC OSCILLATOR

A tuning fork is an example of a damped harmonic oscillator. Indeed we hear the note because some of the energy of oscillation is converted into sound. After it is struck the intensity of the sound, which is proportional to the energy of the tuning fork, steadily decreases. However, the frequency of the note does not change. The ends of the tuning fork make thousands of oscillations before the sound disappears and so we can reasonably assume that the degree of damping is small. We may suspect, therefore, that the frequency of oscillation would not be very different if there were no damping. Thus we infer that the displacement x of an end of the tuning fork is described by a relationship of the form

$$x = (\text{amplitude that reduces with } t) \times \cos \omega t$$

where the angular frequency ω is about but not necessarily the same as would be obtained if there were no damping. We shall assume that the amplitude of oscillation decays exponentially with time. The displacement of an end of the tuning fork will therefore vary according to

$$x = A_0 \exp(-\beta t) \cos \omega t \tag{2.1}$$

where A_0 is the initial value of the amplitude and β is a measure of the degree of damping. The minus sign indicates that the amplitude reduces with time. As we shall see, this expression correctly describes the motion of a damped harmonic oscillator when the degree of damping is small and so the assumptions we have made above are reasonable.

2.2 THE EQUATION OF MOTION FOR A DAMPED HARMONIC OSCILLATOR

An example of a damped harmonic oscillator is shown in Figure 2.1. It is similar to the simple harmonic oscillator described in Section (1.2.2) but now the mass is immersed in a viscous fluid. When an object moves through a viscous fluid it experiences a frictional force. This force dampens the motion: the higher the velocity the greater the frictional force. So as a car travels faster the frictional force increases thereby reducing the fuel economy, while the velocity of a falling raindrop reaches a limiting value because of the frictional force. The damping force F_d acting on the mass in Figure 2.1 is proportional to its velocity v so long as v is not too large, i.e.

$$F_d = -bv \tag{2.2}$$

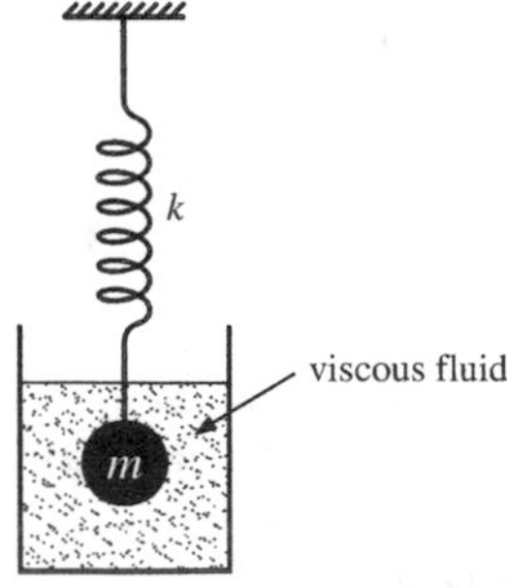

Figure 2.1 An example of a damped mechanical oscillator showing an oscillating mass immersed in a viscous fluid.

where the minus sign indicates that the force always acts in the opposite direction to the motion. The constant b depends on the shape of the mass and the viscosity of the fluid and has the units of force per unit velocity. When the mass is displaced from its equilibrium position there will be the restoring force due to the spring and in addition the damping force $-bv$ due to the fluid. The resulting equation of motion is

$$ma = -kx - bv \tag{2.3a}$$

or

$$m\frac{d^2x}{dt^2} + b\frac{dx}{dt} + kx = 0. \tag{2.3b}$$

We introduce the parameters

$$\omega_o^2 = k/m, \quad \gamma = b/m. \tag{2.4}$$

In terms of these, Equation (2.3b) becomes

$$\boxed{\frac{d^2x}{dt^2} + \gamma\frac{dx}{dt} + \omega_o^2 x = 0.} \tag{2.5}$$

This is the equation of a damped harmonic oscillator. The relationship $k/m = \omega_o^2$ is familiar from our discussion of the simple harmonic oscillator. Now we designate this angular frequency ω_o and describe it as the *natural frequency of oscillation*, i.e. the oscillation frequency if there were no damping. This allows the possibility that the damping does change the frequency of oscillation. In the present example the damping force is linearly proportional to velocity. This linear dependence is very convenient as it has led to an equation that we can readily solve. A damping force proportional to, say, v^2 would be much more difficult to handle. Fortunately, this linear dependence is a good approximation for many other oscillating systems when the velocity is small. Equation (2.5) has different solutions depending on the degree of damping involved, corresponding to the cases of (i) *light damping*, (ii) *heavy* or *over damping* and (iii) *critical damping*. Light damping is the most important case for us because it involves oscillatory motion whereas the other two cases do not.

2.2.1 Light damping

This condition corresponds to the mass in Figure 2.1 being immersed in a fluid of low viscosity like thin oil or even just air. In our previous, qualitative discussion of a lightly damped oscillator, Section 2.1, we suggested an expression for the displacement that had the form $x = A_0 \exp(-\beta t)\cos\omega t$. We adopt a similar functional form here. Then

$$\frac{dx}{dt} = -A_0 \exp(-\beta t)(\omega \sin\omega t + \beta\cos\omega t)$$

and

$$\frac{d^2x}{dt^2} = A_0 \exp(-\beta t)[2\beta\omega \sin \omega t + (\beta^2 - \omega^2) \cos \omega t].$$

Substituting these into Equation (2.5) and collecting terms in $\sin \omega t$ and $\cos \omega t$ gives

$$A_0 \exp(-\beta t)[(2\beta\omega - \gamma\omega) \sin \omega t + (\beta^2 - \omega^2 - \gamma\beta + \omega_o^2) \cos \omega t] = 0.$$

This can only be true for all times if the $\sin \omega t$ and $\cos \omega t$ terms are both equal to zero. Therefore,

$$2\beta\omega - \gamma\omega = 0$$

giving $\beta = \gamma/2$ and

$$\beta^2 - \omega^2 - \gamma\beta + \omega_o^2 = 0.$$

Substituting for β we obtain

$$\omega^2 = \omega_o^2 - \gamma^2/4. \tag{2.6}$$

So our solution for the equation of the lightly damped oscillator is

$$\boxed{x = A_0 \exp(-\gamma t/2) \cos \omega t} \tag{2.7}$$

where $\omega = (\omega_o^2 - \gamma^2/4)^{1/2}$. Equation (2.7) represents oscillatory motion if ω is real, i.e. $\gamma^2/4 < \omega_o^2$ is the condition for light damping. Equation (2.6) shows that the angular frequency of oscillation ω is approximately equal to the undamped value ω_o when $\gamma^2/4 \ll \omega_o^2$. To obtain the general solution of Equation (2.5) we need to include a phase angle ϕ giving

$$x = A_0 \exp(-\gamma t/2) \cos(\omega t + \phi). \tag{2.8}$$

The parameters γ and ω are determined solely by the properties of the oscillator while the constants A_0 and ϕ are determined by the initial conditions. For convenience in our following discussion we will take $\phi = 0$. If we let $\gamma = 0$ we obtain, as expected, our previous results for the simple harmonic oscillator.

A graph of $x = A_0 \exp(-\gamma t/2) \cos \omega t$ is shown in Figure 2.2 where the steady decrease in the amplitude of oscillation is apparent. The dotted lines represent the $\exp(-\gamma t/2)$ term which forms an *envelope* for the oscillations. The zeros in x occur when $\cos \omega t$ is zero and so are separated by π/ω. Therefore the period of the oscillation T, equal to twice this separation, is $2\pi/\omega$. Successive maxima are also separated by T. We consider successive maxima A_n and A_{n+1}. If A_n occurs at time t_o then

$$A_n = x(t_o) = A_0 \exp(-\gamma t_o/2) \cos \omega t_o$$

and

$$A_{n+1} = x(t_o + T) = A_0 \exp[-\gamma(t_o + T)/2] \cos \omega(t_o + T).$$

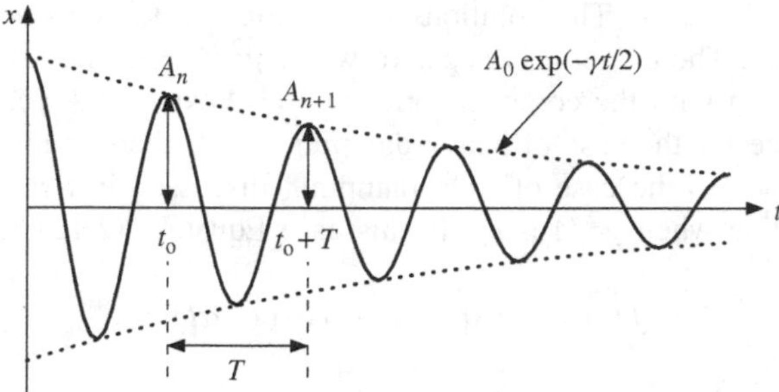

Figure 2.2 A graph of $x = A_0 \exp(-\gamma t/2)\cos\omega t$ illustrating the decay in amplitude of a damped harmonic oscillator. The dotted lines represent the $\exp(-\gamma t/2)$ term of Equation (2.8), which forms an *envelope* of the oscillations.

Since $\cos\omega t_o = \cos\omega(t_o + T)$ we have

$$\frac{A_n}{A_{n+1}} = \exp\left(\frac{\gamma T}{2}\right). \tag{2.9}$$

We see that successive maxima decrease by the same fractional amount. The natural logarithm of A_n/A_{n+1}, i.e.

$$\ln\left(\frac{A_n}{A_{n+1}}\right) = \frac{\gamma T}{2},$$

is called the *logarithmic decrement* and is a measure of this decrease. Note that the larger amplitude occurs in the numerator of this expression.

2.2.2 Heavy damping

Heavy damping occurs when the degree of damping is sufficiently large that the system returns sluggishly to its equilibrium position without making any oscillations at all. This corresponds to the mass in Figure 2.1 being immersed in a fluid of large viscosity like syrup. For this case the oscillatory part of our solution, $\cos\omega t$ in Equation (2.1), is no longer appropriate. Instead we replace it with the general function $f(t)$, i.e.

$$x = \exp(-\beta t) f(t). \tag{2.10}$$

Substituting x and its derivatives into Equation (2.5) and letting $\beta = \gamma/2$ gives

$$\frac{d^2 f}{dt^2} + (\omega_o^2 - \gamma^2/4) f = 0 \tag{2.11}$$

or

$$\frac{d^2 f}{dt^2} = \alpha^2 f \tag{2.12}$$

where $\alpha^2 = (\gamma^2/4 - \omega_0^2)$. The solutions to Equation (2.12) depend dramatically on the sign of α^2. The α^2 term is negative when $\gamma^2/4 < \omega_0^2$ and this leads to an oscillatory solution with the complex form $f(t) = A \exp i(\alpha t + \phi)$. This solution is not appropriate for the case of heavy damping where there is no oscillation. In fact it corresponds to the case of light damping, discussed in Section 2.2.1. The α^2 term is positive when $\gamma^2/4 > \omega_0^2$. In this case Equation (2.12) has the general solution

$$f(t) = A \exp(\alpha t) + B \exp(-\alpha t),$$

giving

$$\begin{aligned} x &= \exp(-\gamma t/2)[A \exp(\alpha t) + B \exp(-\alpha t)] \\ &= A \exp[-\gamma/2 + (\gamma^2/4 - \omega_0^2)^{1/2}]t + B \exp[-\gamma/2 - (\gamma^2/4 - \omega_0^2)^{1/2}]t. \end{aligned} \quad (2.13)$$

This is the non-oscillatory solution that we require. The term $(\gamma^2/4 - \omega_0^2)^{1/2}$ is clearly less than $\gamma/2$ and so the exponents of both exponential terms are negative in sign. Hence the displacement reduces to zero with time and there is no oscillation.

2.2.3 Critical damping

An interesting situation occurs when $\gamma^2/4 = \omega_0^2$. Then Equation (2.12) becomes

$$\frac{\mathrm{d}^2 f}{\mathrm{d}t^2} = 0. \quad (2.14)$$

This equation has the general solution

$$f = A + Bt, \quad (2.15)$$

leading to

$$x = A \exp(-\gamma t/2) + Bt \exp(-\gamma t/2) \quad (2.16)$$

where A has the dimension of length and B has the dimensions of velocity. This is the case of critical damping. Here the mass returns to its equilibrium position in the shortest possible time without oscillating. Critical damping has many important practical applications. For example, a spring may be fitted to a door to return it to its closed position after it has been opened. In practice, however, critical damping is applied to the spring mechanism so that the door returns quickly to its closed position without oscillating. Similarly, critical damping is applied to analogue meters for electrical measurements. This ensures that the needle of the meter moves smoothly to its final position without oscillating or overshooting so that a rapid reading can be taken. Springs are used in motor cars to provide a smooth ride. However damping is also applied in the form of shock absorbers as illustrated schematically in Figure 2.3. Without these the car would continue to bounce up and down long after it went over a bump in the road. A shock

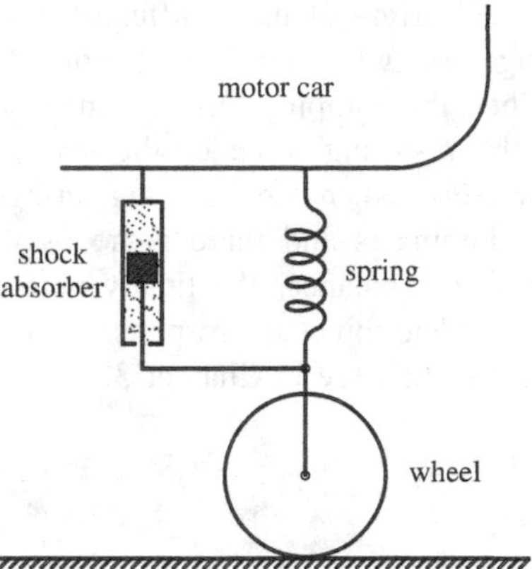

Figure 2.3 Schematic diagram of a car suspension system showing the spring and shock absorber.

absorber consists essentially of a piston that moves in a cylinder containing a viscous fluid. Holes in the piston allow it to move up and down in a damped manner and the damping constant is adjusted so that the suspension system is close to the condition of critical damping. You can see the effect of a shock absorber by pushing down on the front of a car, just above a wheel. The car quickly returns to equilibrium with little or no oscillation. You may also notice that the resistance is greater when you push down quickly than when you push down slowly. This reflects the dependence of the damping force on velocity.

In summary we find three types of damped motion and these are illustrated in Figure 2.4. They correspond to the conditions:

(i) $(\gamma^2/4 < \omega_0^2)$ Light damping; damped oscillations.

(ii) $(\gamma^2/4 > \omega_0^2)$ Heavy damping; exponential decay of displacement.

(iii) $(\gamma^2/4 = \omega_0^2)$ Critical damping; quickest return to equilibrium position without oscillation.

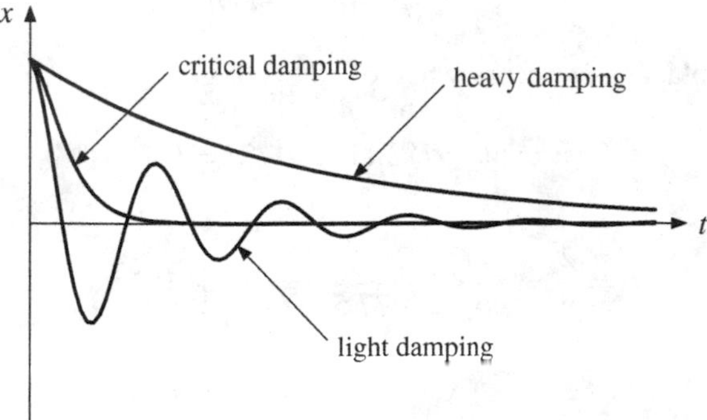

Figure 2.4 The motion of a damped oscillator for the cases of light damping, heavy damping and critical damping.

To appreciate the physical origin of these different types of motion, we recall that $\gamma^2/4$ is the damping term while ω_o^2 is proportional to the spring constant k through $\omega_o^2 = k/m$. When the damping term is small compared with k/m, the motion is governed by the restoring force of the spring and we have damped oscillatory motion. Conversely, when the damping term is large compared with k/m the damping force dominates and there is no oscillation. At the point of critical damping the two forces balance. We finally note that the relative size of $\gamma^2/4$ compared with ω_o^2 also determines the response of the oscillator to an applied periodic driving force, as we shall see in Chapter 3.

Worked example

A mass of 2.5 kg is attached to a spring that has a value of k equal to 600 N m^{-1}. (a) Determine the value of the damping constant b that is required to produce critical damping. (b) The mass receives an impulse that gives it a velocity of $v_i = 1.5$ m s^{-1} at $t = 0$. Determine the maximum value of the resultant displacement and the time at which this occurs.

Solution

(a) For critical damping, $\gamma^2/4 = b^2/4m^2 = \omega_o^2 = k/m$. Therefore,

$$b = \sqrt{4mk} = \sqrt{4 \times 2.5 \times 600} = 77.5 \text{ kg s}^{-1}.$$

(b) General solution for critical damping is

$$x = A\exp(-\gamma t/2) + Bt\exp(-\gamma t/2).$$

Therefore

$$v = \frac{\mathrm{d}x}{\mathrm{d}t} = \exp(-\gamma t/2)(B - \gamma Bt/2 - \gamma A/2).$$

Initial conditions, $x = 0$ and $v = v_i$ at $t = 0$, give $A = 0$ and $B = v_i$. Therefore,

$$x(t) = v_i t\exp(-\gamma t/2).$$

Maximum displacement occurs when $\mathrm{d}x/\mathrm{d}t = 0$, giving

$$v_i\exp(-\gamma t/2)(1 - \gamma t/2) = 0.$$

Hence

$$t = \frac{2}{\gamma} = \frac{2m}{b} = \frac{2 \times 2.5}{77.5} = 6.5 \times 10^{-2} \text{ s}$$

and

$$x = \frac{2v_i}{\mathrm{e}\gamma} = \frac{2mv_i}{\mathrm{e}b} = \frac{2 \times 2.5 \times 1.5}{\mathrm{e} \times 77.5} = 3.6 \times 10^{-2} \text{ m}.$$

x is a product of a linearly increasing function and an exponentially decaying one. Of course the exponential function wins in the end and the displacement steadily reduces to zero as shown in Figure 2.5.

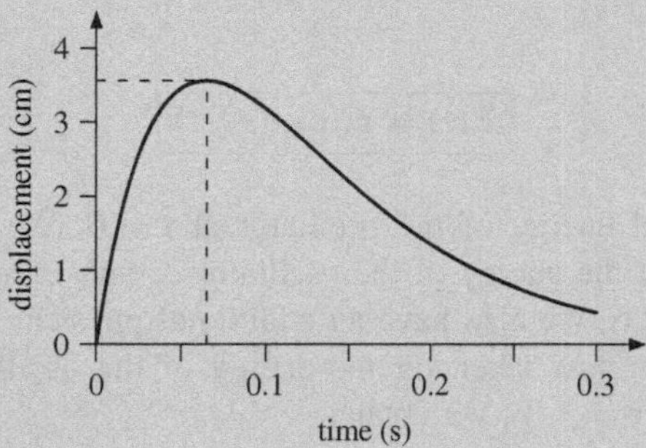

Figure 2.5 An example of critical damping showing the steady decrease of the displacement to zero following an impulse applied at time $t = 0$. The dashed lines indicate the maximum value of the displacement and the time at which it occurs.

2.3 RATE OF ENERGY LOSS IN A DAMPED HARMONIC OSCILLATOR

The mechanical energy of a damped harmonic oscillator is eventually dissipated as heat to its surroundings. We can deduce the rate at which energy is lost by considering how the total energy of the oscillator changes with time. The total energy E is given by

$$E = K + U = \frac{1}{2}mv^2 + \frac{1}{2}kx^2. \tag{1.19}$$

For the case of a very lightly damped oscillator, i.e. $\gamma^2/4 \ll \omega_o^2$, it follows from Equation (2.6) that to a good approximation $\omega = \omega_o$ and from Equation (2.7) that

$$x = A_0 \exp(-\gamma t/2) \cos \omega_o t. \tag{2.17}$$

Hence,

$$v = \frac{\mathrm{d}x}{\mathrm{d}t} = -A_0\omega_o \exp(-\gamma t/2)[\sin \omega_o t + (\gamma/2\omega_o) \cos \omega_o t].$$

Since $\gamma/2 \ll \omega_o$, we can neglect the second term in the square brackets and write

$$v = \frac{\mathrm{d}x}{\mathrm{d}t} = -A_0\omega_o \exp(-\gamma t/2)(\sin \omega_o t).$$

Then

$$E = \frac{1}{2}A_0^2 \exp(-\gamma t)(m\omega_o^2 \sin^2 \omega_o t + k \cos^2 \omega_o t).$$

Substituting for $\omega_0^2 = k/m$, we obtain

$$E = \frac{1}{2}kA_0^2 \exp(-\gamma t)$$

giving

$$\boxed{E(t) = E_0 \exp(-\gamma t)} \tag{2.18}$$

where E_0 is the total energy of the oscillator at $t = 0$. We have the important and simple result that the energy of the oscillator decays exponentially with time as shown in Figure 2.6. We also have an additional physical meaning for γ. The reciprocal of γ is the time taken for the energy of the oscillator to reduce by a factor of e. Defining $\tau = 1/\gamma$, we obtain

$$E(t) = E_0 \exp(-t/\tau) \tag{2.19}$$

where τ has the dimensions of time and is called the *decay time* or *time constant* of the system. There are many examples of both classical and quantum-mechanical systems that give rise to exponential decay of their energy with time as described by Equation (2.19) and for some of these τ is called the *lifetime*.

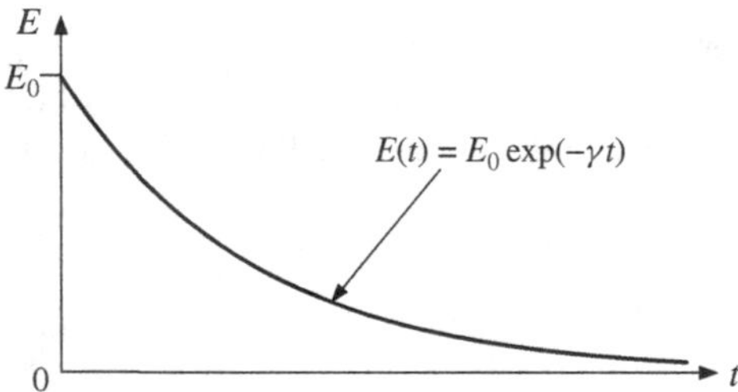

Figure 2.6 The exponential decay of the energy of a very lightly damped harmonic oscillator.

The energy of an oscillator is dissipated because it does work against the damping force at the rate (damping force × velocity). We can see this by differentiating Equation (1.19) with respect to time. Thus

$$\frac{\mathrm{d}E}{\mathrm{d}t} = \frac{\mathrm{d}}{\mathrm{d}t}\left(\frac{1}{2}mv^2 + \frac{1}{2}kx^2\right) = mv\frac{\mathrm{d}v}{\mathrm{d}t} + kx\frac{\mathrm{d}x}{\mathrm{d}t} = (ma + kx)v$$

which, using Equation (2.3a), gives

$$\frac{\mathrm{d}E}{\mathrm{d}t} = (-bv)v. \tag{2.20}$$

2.3.1 The quality factor Q of a damped harmonic oscillator

It is useful to have a *figure of merit* to describe how good an oscillator is, where we imply that the smaller the degree of damping the higher the *quality* of the oscillator. Moreover we would like a figure of merit that is dimensionless and can readily be applied to any oscillator whether it is mechanical, electrical or otherwise. This is called the *quality factor* Q of the oscillator and is defined below. It is reasonable to expect that an oscillator with a high Q-value would make an appreciable number of oscillations before its energy is reduced substantially, say by a factor of e. Equation (2.19) shows that this reduction occurs after time τ and we might therefore compare τ with the period of oscillation T. If τ is large compared with T we would have many oscillations and the Q-value of the oscillator would be large. Conversely when τ approaches T in value there would be a small number of oscillations and the Q-value would be small. Thus the ratio τ/T would provide us with a useful figure of merit. Conventionally, however, it is quantities that are related to the inverse of τ and T that are compared. These are the damping term γ and the angular frequency of oscillation which is equal to ω_0 to a very good approximation under most circumstances. The quality factor Q is therefore defined as

$$\boxed{Q = \frac{\omega_0}{\gamma}.} \tag{2.21}$$

γ and ω_0 have the same dimensions as each other, $[\text{time}]^{-1}$, and Q is a pure, dimensionless number. In Section 2.2 we compared the relative sizes of $\gamma^2/4$ and ω_0^2 to deduce what sort of damped motion would result. We now have a new physical interpretation for this comparison. The reciprocal of γ is a characteristic time for the exponential decay of the energy and the reciprocal of ω_0 is a characteristic of the oscillation period. Figure 2.7 shows the behaviour of a particular oscillator with various amounts of applied damping together with the respective Q-values. It is quite evident that the higher the Q-value, the greater the number of oscillations. Also shown for comparison is the behaviour of the oscillator for the condition of critical damping.

We can define the quality factor Q in a different way by considering the rate at which the energy of the oscillator is dissipated. If we consider the energy of a very lightly damped oscillator one period apart we have from Equation (2.18),

$$E_1 = E_0 \exp(-\gamma t), \quad E_2 = E_0 \exp[-\gamma(t+T)]$$

giving

$$\frac{E_2}{E_1} = \exp(-\gamma T).$$

The series expansion of e^x is

$$e^x = 1 + x + \frac{x^2}{2!} + \frac{x^3}{3!} \cdots.$$

Thus, when $x \ll 1$,

$$e^x \simeq 1 + x$$

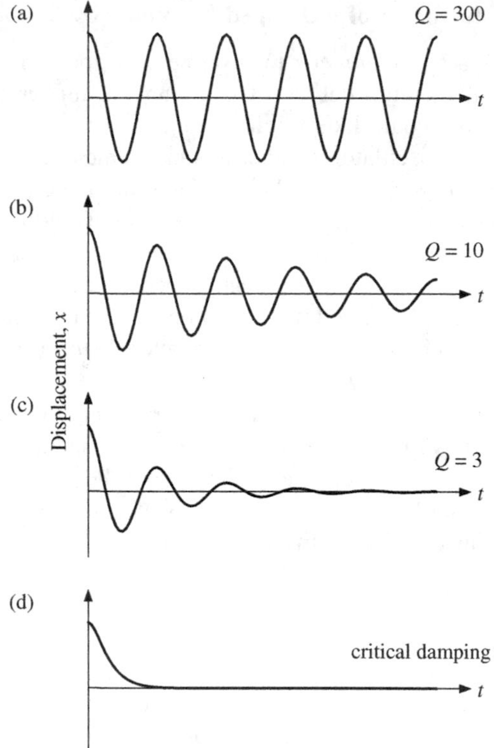

Figure 2.7 The behaviour of an oscillator with various degrees of damping. The corresponding Q-values are (a) 300, (b) 10 and (c) 3. The case of critical damping for the oscillator is also shown (d).

to a good approximation. For a very lightly damped oscillator, we have $\gamma T \ll 1$ and therefore

$$\frac{E_1 - E_2}{E_1} \simeq \gamma T \simeq \frac{2\pi\gamma}{\omega_o} = \frac{2\pi}{Q} \tag{2.22}$$

where we have substituted ω_o for ω. The fractional change in energy per cycle is equal to $2\pi/Q$ and so the fractional change in energy per radian is equal to $1/Q$. We then define Q by

$$Q = \frac{\text{energy stored in the oscillator}}{\text{energy dissipated per radian}}. \tag{2.23}$$

We can usefully cast our previous equations in terms of the dimensionless quantity Q. Thus the equation of a damped oscillator, Equation (2.5), becomes

$$\frac{d^2x}{dt^2} + \frac{\omega_o}{Q}\frac{dx}{dt} + \omega_o^2 x = 0 \tag{2.24}$$

and the angular frequency ω, Equation (2.6), becomes

$$\omega = \omega_o(1 - 1/4Q^2)^{1/2}. \tag{2.25}$$

Equation (2.25) confirms our assumption that ω is equal to ω_o to a good approximation under most circumstances. Even when Q is as low as 5, ω is different from ω_o by just 0.5%.

Worked example

When the E string of a guitar (frequency 330 Hz) is plucked, the sound intensity decreases by a factor of 2 after 4 s. Determine (i) the decay time τ, (ii) the quality factor Q and (iii) the fractional energy loss per cycle.

Solution

(i) The sound intensity is proportional to the energy of oscillation.

$$E(t) = E_0 \exp(-t/\tau)$$

giving

$$\tau = \frac{t}{\ln[E_0/E(t)]} = \frac{4}{\ln 2} = 5.77 \text{ s.}$$

(ii)

$$Q = \omega_o/\gamma = \omega_o\tau = 2\pi \times 330 \times 5.77 = 1.2 \times 10^4.$$

(iii)

$$\frac{\Delta E}{E} = \frac{2\pi}{Q} = 5.3 \times 10^{-4}.$$

Worked example

The electron in an excited atom behaves like a damped harmonic oscillator when the atom radiates light. The lifetime of an excited atomic state is 10^{-8} s and the wavelength of the emitted light is 500 nm. Deduce a value for the quality factor.

Solution

The lifetime corresponds to τ which is equal to $1/\gamma$. The frequency of oscillation ν is given by $\nu = c/\lambda$ and so $\omega_o = 2\pi c/\lambda$. Therefore

$$Q = \frac{\omega_o}{\gamma} = \frac{2\pi \times 3 \times 10^8 \times 10^{-8}}{500 \times 10^{-9}} \approx 4 \times 10^7$$

which is a very high value indeed.

TABLE 2.1 Typical values of Q for a variety of damped oscillators.

Damped oscillator system	Typical value of Q
Paper weight suspended on a rubber band	10
Clock pendulum	75
Electrical LCR circuit	200
Plucked violin string	10^3
Microwave cavity oscillator	10^4
Quartz crystal	10^6

Typical values of Q for a variety of damped oscillators are presented in Table 2.1.

2.4 DAMPED ELECTRICAL OSCILLATIONS

In our mechanical example of a mass moving through a fluid we saw that the fluid offered a resistance that damped the motion. In the case of an electrical oscillator it is the resistance in the circuit that impedes the flow of current. An electrical oscillator is shown in Figure 2.8. It consists of an inductor L and capacitor C

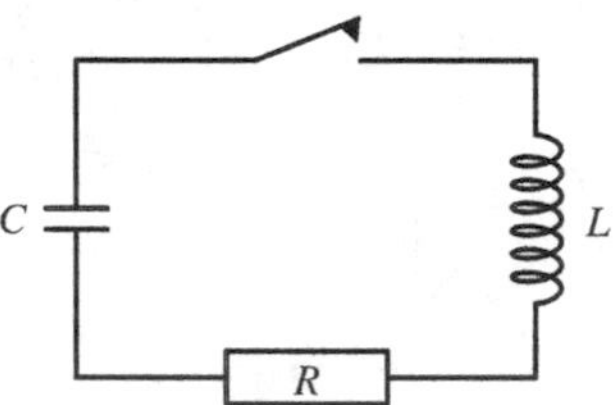

Figure 2.8 The circuit of a damped electrical oscillator consisting of an inductor L, a capacitor C and a resistor R connected in series.

as before (see Figure 1.21) but now there is also the resistor R. We charge the capacitor to voltage $V_C = q/C$, and then close the switch. Kirchoff's law gives

$$L\frac{\mathrm{d}I}{\mathrm{d}t} + RI + \frac{q}{C} = 0$$

or

$$L\frac{\mathrm{d}^2q}{\mathrm{d}t^2} + R\frac{\mathrm{d}q}{\mathrm{d}t} + \frac{q}{C} = 0. \tag{2.26}$$

This has the identical form to Equation (2.3b),

$$m\frac{\mathrm{d}^2x}{\mathrm{d}t^2} + b\frac{\mathrm{d}x}{\mathrm{d}t} + kx = 0, \tag{2.3b}$$

and we recognise the analogous quantities we met before: q is equivalent to x, L to m and k to $1/C$. However, we see that R is analogous to the mechanical damping constant b and so R/L is the equivalent of γ $(= b/m)$. Since the above differential equations have identical forms, their solutions also have identical forms. The importance of this is that we can use our results for the mechanical oscillator to immediately write down the corresponding results for the electrical case. Thus from Equation (2.7) it follows that

$$q = q_0 \exp(-Rt/2L) \cos[(1/LC - R^2/4L^2)^{1/2} t] \tag{2.27}$$

where q_0 is the initial charge on the capacitor. This corresponds to the case of light damping which now means that $R^2/4L^2 < 1/LC$. Since the voltage V_C across the capacitor is equal to q/C

$$V_C = V_0 \exp(-Rt/2L) \cos[(1/LC - R^2/4L^2)^{1/2} t] \tag{2.28}$$

where V_0 is the initial value of the voltage. This is an oscillating voltage at an angular frequency ω given by

$$\omega^2 = \frac{1}{LC} - \frac{R^2}{4L^2} \tag{2.29}$$

which is essentially equal to $1/LC$ when $R^2/4L^2 \ll 1/LC$. The amplitude of the oscillating voltage decays exponentially with a time constant of $R/2L$ and so R/L has the dimensions of $[\text{time}]^{-1}$. For $R^2/4L^2 > 1/LC$ we have the case of heavy damping and for $R^2/4L^2 = 1/LC$ we have critical damping. Similarly we find that the quality factor Q of the circuit is given by

$$Q = \frac{1}{R}\sqrt{\frac{L}{C}}. \tag{2.30}$$

For example, with $L = 10$ mH, $C = 2.5$ nF and $R = 10\,\Omega$, $Q = 200$, which is a typical value for an electrical oscillator. Again we emphasise the exact correspondence between the equations and solutions that describe the mechanical and electrical systems, so that mechanical systems can be simulated by electrical circuits. Such analogue computers can greatly facilitate the design and development of mechanical systems.

PROBLEMS 2

2.1 A spring balance consists of a pan that hangs from a spring. A damping force $F_d = -bv$ is applied to the balance so that when an object is placed in the pan it comes to rest in the minimum time without overshoot. Determine the required value of b for an object of mass 2.5 kg that extends the spring by 6.0 cm. (Assume $g = 9.81$ m s^{-2}.)

2.2 A mass of 0.30 kg hangs from the end of a light spring. The system is damped by a light sail attached to the mass so that the ratio of amplitudes of consecutive oscillations is equal to 0.90. It is found that 10 complete oscillations takes 25 s. Obtain a quantitative expression for the damping force and determine the damping factor γ of the system.

2.3

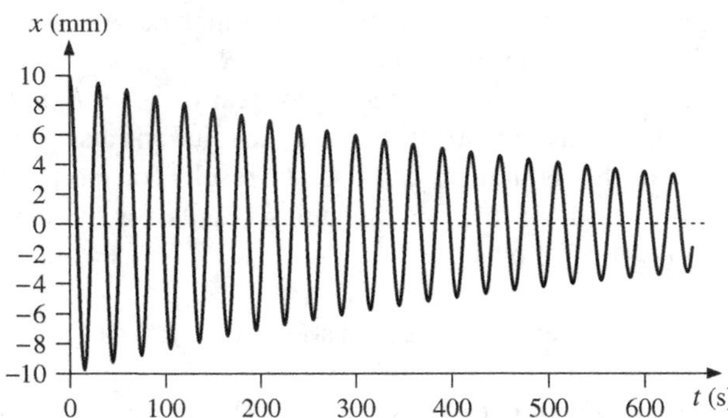

The figure shows a graph of displacement x against time t for a damped harmonic oscillator. Deduce the quality factor Q of the oscillator.

2.4 The energy of a damped harmonic oscillator is observed to reduce by a factor of 2 after 10 complete cycles. By what factor will it reduce after 50 complete cycles?

2.5 An undamped oscillator has a natural frequency ω_o of π rad s^{-1}. Various amounts of damping are added to the system to give values of the damping factor γ equal to 0.01, 0.30 and 1.0 s^{-1}, respectively. (a) For each value of γ find the corresponding Q-value and frequency ω of the damped oscillations. Comment on the change in ω over this range of γ. (b) For each of the values of Q, use a spreadsheet program to plot $x = A_0 \exp(-\gamma t/2) \cos \omega t$ over the time period $t = 0$ to 10 s, using a value of 10 mm for A_0. (c) Obtain an expression for x for the case of critical damping with the initial conditions, $x = 10$ mm and $dx/dt = 0$. Plot x over the time period $t = 0$ to 10 s.

2.6 When damping is applied to a simple harmonic oscillator its frequency of oscillation changes from ω_o to a different frequency ω. Show for a very lightly damped harmonic oscillator of quality factor Q that the fractional change in frequency is equal to $1/8Q^2$ to a good approximation.

2.7 A simple pendulum is constructed from an aluminium sphere attached to a light rod. A second pendulum is constructed of the same length but with a brass sphere. The diameters of the two spheres are the same. The two pendulums are set in motion at the same time with the same amplitude of oscillation. After 10 min the amplitude of oscillation of the aluminium pendulum has decreased to one-half its initial value. By what factor has the amplitude of oscillation of the brass pendulum decreased at this time? (Assume that the damping force acting on a pendulum is directly proportional to the velocity of the sphere. The densities of aluminium and brass are 2.7×10^3 kg m^{-3} and 8.5×10^3 kg m^{-3}, respectively.)

2.8 According to classical electromagnetic theory, an accelerating electron radiates energy at a rate Ke^2a^2/c^3, where a is the acceleration, e is the electronic charge, c is the velocity of light and K is a constant with a value of 6×10^9 N m^2 C^{-2}. Suppose that the motion of the electron can be represented by the expression $x = A \sin \omega t$ during one cycle of its motion. (a) Show that the energy radiated during one cycle is $Ke^2\pi\omega^3A^2/c^3$. (b) Recalling that the total energy of a harmonic oscillator is $\frac{1}{2}m\omega^2A^2$ where m is the mass, show that the quality factor Q is $mc^3/Ke^2\omega$. (c) Using a typical value of ω for a visible photon, estimate the 'lifetime' of the radiating system ($e = 1.6 \times 10^{-19}$ C, mass of electron $= 9.1 \times 10^{-31}$ kg).

3

Forced Oscillations

So far we have considered *free oscillations* where a system is disturbed from rest and then oscillates about its equilibrium position with steadily decreasing amplitude, as when we strike a bell. We now turn our attention to *forced oscillations* where we apply a periodic driving force to the system. We are surrounded by examples of such forced oscillations. We give a push to a playground swing at regular intervals to sustain its motion. In a pendulum clock the escapement mechanism gives regular impulses to the pendulum and in an analogous fashion the crystal in a crystal-controlled clock receives regular electrical impulses to maintain its oscillation. A musician uses a bow to play a note on a violin while air is driven into the pipes of an organ to sustain a note. (By contrast a harp and a guitar are plucked instruments and provide examples of free oscillations.) On a much larger scale the moon exerts a gravitational pull that exerts a periodic driving force on the oceans of the Earth that strongly influences their tidal motion. At the microscopic level, the radiation in a microwave oven drives the electrons of the water molecules in the item being cooked.

Forced oscillations are the subject of the present chapter. We will see that the system *always* oscillates at the frequency of the applied force, apart from an initial *transient response*. We will see that the frequency of the applied force has a dramatic effect on the amplitude of the oscillations, especially close to the natural frequency of the system. For example, a singer can cause a wine glass to shatter when they produce a note that is at the *resonance* frequency of the glass (the frequency you hear when you tap the glass). The shaking of the ground in an earthquake may cause a building to collapse. The important point is that a periodic force can produce large and possibly catastrophic effects when applied at the resonance frequency. We will see that the sharpness of the response to the applied force depends on the quality factor Q of the oscillator. This is the same factor that we encountered in our discussion of damped harmonic motion in Section 2.3.1. In this chapter we also introduce the complex representation of oscillatory motion. We summarise the basic rules for manipulating complex numbers in Section 3.6.1

and then illustrate their use in the description and analysis of oscillatory motion in Sections 3.6.2 and 3.6.3, respectively.

3.1 PHYSICAL CHARACTERISTICS OF FORCED HARMONIC MOTION

We can observe the main physical characteristics of forced harmonic motion using a simple pendulum. We drive the pendulum by moving its point of suspension backwards and forwards harmonically, along a horizontal direction. At very low driving frequencies the pendulum mass closely follows the movement of the point of suspension with them both moving in the same direction as each other, i.e. they have the same amplitude and move in phase. As the driving frequency is increased the amplitude of oscillation increases dramatically and becomes much larger than the movement of the point of suspension. We might rightly suspect that the maximum amplitude occurs when the pendulum is driven close to its *natural frequency of oscillation*. The system is then said to be *in resonance*. We get the largest amplitude at resonance because this is the frequency at which the pendulum 'wants' to oscillate. As the driving frequency is increased further the amplitude of oscillation decreases but perhaps more surprisingly the mass now moves in the *opposite* direction to the point of suspension, although still with the same frequency. At even higher frequencies we reach the situation where the pendulum mass hardly moves at all. This is because it has inertia. The simple pendulum serves as a useful example, but all forced oscillators behave in this manner.

3.2 THE EQUATION OF MOTION OF A FORCED HARMONIC OSCILLATOR

3.2.1 Undamped forced oscillations

We begin with a mass m on a horizontal spring as shown in Figure 3.1. The spring constant is k and the mass moves without friction on a horizontal surface. The displacement x is measured from the equilibrium position of the mass. This system is similar to the one described in Section 1.2.1 but now we imagine that a periodic driving force $F = F_0 \cos \omega t$ is applied to it. The mass is acted upon by the combination of the restoring force from the spring and the applied driving force. From Newton's second law we obtain

$$\boxed{m\frac{\mathrm{d}^2x}{\mathrm{d}t^2} + kx = F_0 \cos \omega t.} \tag{3.1}$$

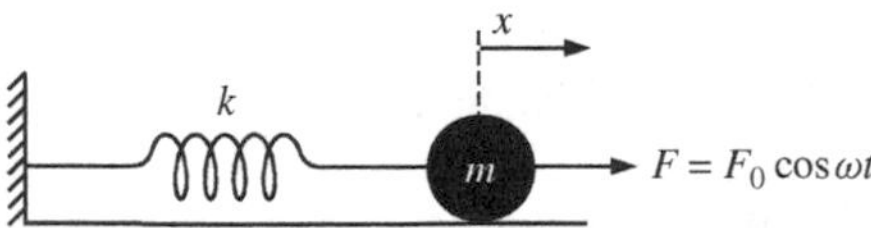

Figure 3.1 Application of a periodic driving force $F = F_0 \cos \omega t$ to a harmonic oscillator consisting of a mass m on the end of a spring of spring constant k.

This is the equation of motion for forced oscillations of a harmonic oscillator when there is no damping. We shall deduce a solution for Equation (3.1) and see how the oscillator behaves as we change the angular frequency ω of the driving force. First we note one limit of ω, namely $\omega = 0$. If we have a force F_0 that does not change with time, i.e. $\omega = 0$, the acceleration term is zero. The displacement x is then equal to F_0/k. So, at very low driving frequencies when ω tends to zero, the amplitude of oscillation will tend to the value F_0/k.

We deduce a solution for forced oscillations, Equation (3.1), using the arrangement of a mass m on a vertical spring, of spring constant k, as shown in Figure 3.2. Here we move the upper end s of the spring up and down harmonically in the vertical direction according to $\xi = a \cos \omega t$ where a is the amplitude and ω is the applied frequency. (This simple but very informative experiment can be performed using a few elastic bands strung together with a small mass attached to the lower end.) We measure the displacement x from the equilibrium position of the mass and take displacements in the downward direction as positive. The resultant equation of motion is

$$m\frac{\mathrm{d}^2x}{\mathrm{d}t^2} = -k(x - \xi) \tag{3.2}$$

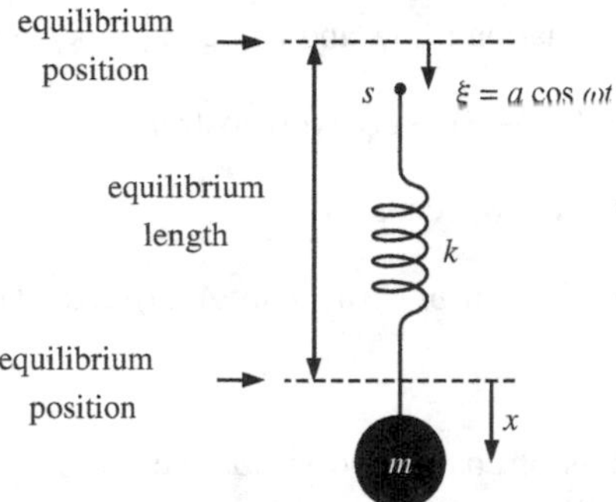

Figure 3.2 Practical realisation of forced oscillations, where the top of the spring s is moved up and down harmonically about its equilibrium position.

or, substituting for ξ,

$$m\frac{\mathrm{d}^2x}{\mathrm{d}t^2} + kx = ka\cos\omega t. \tag{3.3}$$

Equation (3.3) has exactly the same form as Equation (3.1) with

$$ka = F_0. \tag{3.4}$$

The response of the system is similar to that of the driven pendulum described in Section 3.1. At very low frequencies the amplitude of oscillations tends to the value of the amplitude a of the point of suspension. Under these conditions the motion is governed by the spring constant or *stiffness* of the spring. As ω

is increased the amplitude of oscillation increases dramatically as the resonance frequency is approached. As ω is increased above the resonance frequency the amplitude decreases and the mass moves in the opposite direction to the driving force. At higher frequencies still, the amplitude tends to zero when the motion is governed by the inertia of the mass. At all frequencies, however, the mass moves up and down periodically at the same frequency ω as the driving force. This behaviour suggests that the displacement x of the mass can be written as

$$x = A(\omega)\cos(\omega t - \delta). \tag{3.5}$$

In this equation, $A(\omega)$ is the physical amplitude that we observe and which we naturally define as a positive quantity. δ is a phase angle but has a different meaning to the phase angle ϕ in the expression given in Chapter 1:

$$x = A\cos(\omega t + \phi). \tag{1.14}$$

In Equation (1.14) ϕ relates to the initial conditions and along with A completely defines the free oscillations. In Equation (3.5), δ is the phase angle between the driving force and the resultant displacement. The minus sign of δ in Equation (3.5) implies that the displacement lags behind the driving force and this is indeed the case in forced oscillations. From our previous considerations, we expect δ to be zero at very low frequencies and equal to π at very high frequencies. Substituting x and its second derivative in Equation (3.3), and using $\omega_0^2 = k/m$ (Equation (2.4)) gives

$$-\omega^2 A(\omega)\cos(\omega t - \delta) + \omega_0^2 A(\omega)\cos(\omega t - \delta) = \omega_0^2 a\cos\omega t.$$

Expanding terms in $\cos(\omega t - \delta)$ leads to

$$\begin{aligned}&-\omega_0^2 A(\omega)(\cos\omega t\cos\delta + \sin\omega t\sin\delta) + \omega_0^2 A(\omega)(\cos\omega t\cos\delta + \sin\omega t\sin\delta)\\&= \omega_0^2 a\cos\omega t.\end{aligned}$$

Then equating coefficients of $\cos\omega t$ and $\sin\omega t$ we obtain

$$A(\omega)(1 - \omega^2/\omega_0^2)\cos\delta = a \tag{3.6a}$$

and

$$A(\omega)(1 - \omega^2/\omega_0^2)\sin\delta = 0. \tag{3.6b}$$

Dividing Equation (3.6b) by Equation (3.6a) gives $\tan\delta = 0$ and so $\delta = 0$ or π as expected. When $\delta = 0$, Equation (3.6a) gives

$$A(\omega) = \frac{a}{(1 - \omega^2/\omega_0^2)}. \tag{3.7}$$

Since the amplitude $A(\omega)$ is defined as a positive quantity, Equation (3.7) shows that ω must be less than ω_0 when $\delta = 0$. When $\delta = \pi$, Equation (3.6a) gives

$$A(\omega) = \frac{-a}{(1 - \omega^2/\omega_0^2)}. \tag{3.8}$$

Now $A(\omega)$ is a positive quantity only when ω is greater than ω_o. Thus we conclude that $x = A(\omega)\cos(\omega t - \delta)$ is a solution for the undamped forced oscillator and that $\delta = 0$ for $\omega < \omega_o$ and $\delta = \pi$ for $\omega > \omega_o$.

Equation (3.7) shows that $A(\omega)$ tends to $a(= F_0/k)$ as ω tends to zero and Equation (3.8) shows that $A(\omega)$ tends to zero as ω tends to infinity, as we expect. However we also see that $A(\omega)$ tends to infinity as ω approaches ω_o. A plot of $A(\omega)$ against ω, according to Equations (3.7) and (3.8), is shown in Figure 3.3. The behaviour of $A(\omega)$ as ω approaches ω_o is clearly unphysical and arises because damping forces have been neglected. These are present in real systems and limit the maximum value of $A(\omega)$ as we shall see in Section 3.2.2. Figure 3.3 also shows the behaviour of the phase angle δ with respect to ω. The change of the phase angle from zero to π is consistent with the behaviour of the forced oscillators we have considered but its sharp and abrupt change at $\omega = \omega_o$ is unphysical. This is again because the effects of damping have not been included.

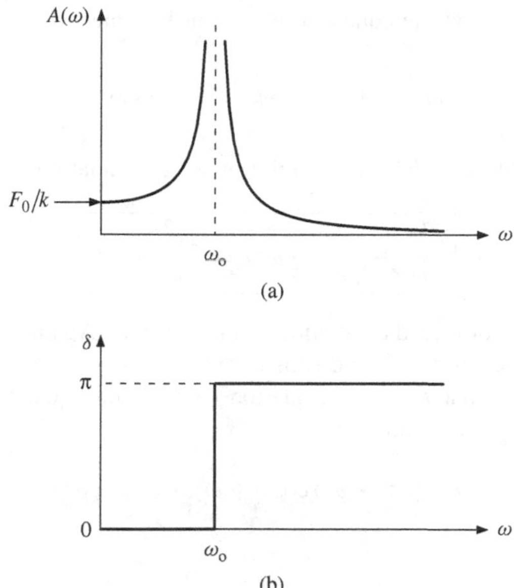

Figure 3.3 (a) A plot of the amplitude of oscillation $A(\omega)$ of a forced oscillator against driving frequency ω, when there is no damping. (b) The variation of the phase angle δ with driving frequency. δ is the phase angle between the driving force and the resultant displacement which lags behind the driving force.

The amplitude $A(\omega)$ which we have defined above is the *physical* amplitude. It is always positive and is given by *different* expressions, (3.7) and (3.8) for $\omega < \omega_o$ and $\omega > \omega_o$, respectively. An alternative description, which avoids this division and so allows both situations to be handled simultaneously, is sometimes convenient (cf. Section 4.5). Instead of Equation (3.5), we write the solution of Equation (3.3) in the form

$$x = C(\omega)\cos\omega t. \tag{3.5a}$$

Comparing Equations (3.5) and (3.5a), and using Equations (3.7) and (3.8), at once gives

$$C(\omega) = \frac{a}{(1 - \omega^2/\omega_0^2)} \tag{3.7a}$$

for *both* $\omega < \omega_0$ and $\omega > \omega_0$. [For $\omega > \omega_0$, this follows since $\delta = \pi$ and $A(\omega)\cos(\omega t - \pi) = -A(\omega)\cos\omega t$.] We might call $C(\omega)$ the *algebraic* amplitude which, in contrast to the *physical* amplitude $A(\omega)$, is given by the same expression for *all* values of ω. In contrast to $A(\omega)$, which is always positive, $C(\omega)$ is positive for $\omega < \omega_0$ and negative for $\omega > \omega_0$.

3.2.2 Forced oscillations with damping

We will again assume that the damping force is directly proportional to the velocity of the mass as we did in Section 2.2. This adds the damping term $b\mathrm{d}x/\mathrm{d}t$ to Equation (3.1), so that the equation of motion becomes

$$m\frac{\mathrm{d}^2x}{\mathrm{d}t^2} + b\frac{\mathrm{d}x}{\mathrm{d}t} + kx = F_0\cos\omega t. \tag{3.9}$$

We make the substitutions $b/m = \gamma$ and $k/m = \omega_0^2$, Equation (2.4), to obtain

$$\boxed{\frac{\mathrm{d}^2x}{\mathrm{d}t^2} + \gamma\frac{\mathrm{d}x}{\mathrm{d}t} + \omega_0^2x = \frac{F_0}{m}\cos\omega t.} \tag{3.10}$$

This is the equation for forced oscillations with damping. Again we try a solution of the form $x = A(\omega)\cos(\omega t - \delta)$ and substitute for x and its derivatives in Equation (3.10), remembering that $F_0 = ka$, Equation (3.4). Then equating the coefficients of $\cos\omega t$ and $\sin\omega t$ we obtain

$$A(\omega)[(\omega_0^2 - \omega^2)\cos\delta + \omega\gamma\sin\delta] = \omega_0^2a \tag{3.11a}$$

and

$$(\omega_0^2 - \omega^2)\sin\delta = \omega\gamma\cos\delta, \tag{3.11b}$$

giving

$$\tan\delta = \frac{\omega\gamma}{(\omega_0^2 - \omega^2)}. \tag{3.12}$$

We see that the phase angle δ, as well as the amplitude $A(\omega)$, depends on the driving frequency ω. Using the mathematical nomenclature $\rightarrow$ meaning 'tends to', inspection of Equation (3.12) shows that

as	$\omega \rightarrow 0$,	$\tan\delta \rightarrow 0$,	and	$\delta \rightarrow 0$,
as	$\omega \rightarrow \infty$,	$\tan\delta \rightarrow 1/(-\infty)$,	and	$\delta \rightarrow \pi$,
and when	$\omega = \omega_0$,	$\tan\delta = \infty$,	and	$\delta = \pi/2$

As the angular frequency of the applied force varies from very low to very high values, so tan δ varies *continuously* from zero to π and passes through $\pi/2$ at precisely the frequency ω_o. It may seem surprising that the displacement lags behind the driving force by $\pi/2$ at resonance. However, in a harmonic oscillator, the velocity is always $\pi/2$ ahead of the displacement. This means that at resonance the mass is always moving in the same direction as the driving force, as when we give a push to a playground swing. From Equation (3.12) we can construct the right-angled triangle shown in Figure 3.4 to obtain

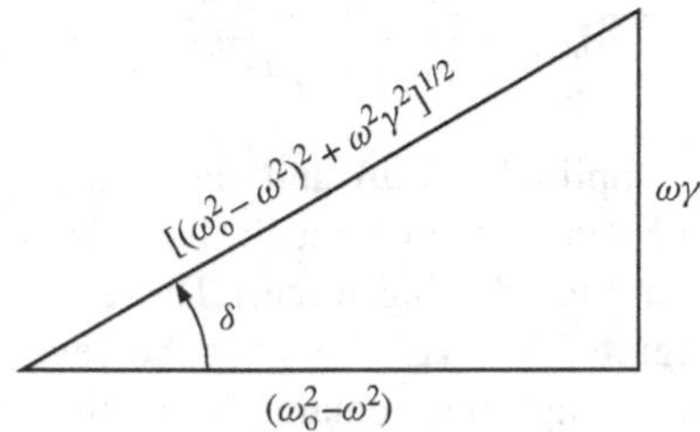

Figure 3.4 Geometrical construction for the phase angle δ.

$$\sin\delta = \frac{\omega\gamma}{[(\omega_o^2-\omega^2)^2+\omega^2\gamma^2]^{1/2}} \tag{3.13}$$

and

$$\cos\delta = \frac{(\omega_o^2-\omega^2)}{[(\omega_o^2-\omega^2)^2+\omega^2\gamma^2]^{1/2}}. \tag{3.14}$$

Substituting for $\sin\delta$ and $\cos\delta$ in Equation (3.11) we finally obtain

$$A(\omega) = \frac{a\omega_o^2}{[(\omega_o^2-\omega^2)^2+\omega^2\gamma^2]^{1/2}} \tag{3.15}$$

which describes the amplitude dependence on driving frequency ω for forced oscillations with damping. We note that Equation (3.15) reduces to the result for the undamped case, when γ is zero. Inspection of Equation (3.15) shows that

as	$\omega \to 0$,	$A(\omega) \to a(=F_0/k)$,
as	$\omega \to \infty$,	$A(\omega) \to 0$,
and when	$\omega = \omega_o$,	$A(\omega) = a\omega_o/\gamma$.

These results are similar to the undamped case except that the amplitude does not go to infinity at $\omega = \omega_o$. Furthermore, the maximum amplitude of oscillation no longer occurs at ω_o. For $A(\omega)$ to be a maximum, the denominator in Equation (3.15) must be a minimum. This occurs when

$$\frac{\mathrm{d}}{\mathrm{d}\omega}[(\omega_o^2-\omega^2)^2+\omega^2\gamma^2]^{1/2} = 0,$$

from which

$$\omega = \omega_o(1 - \gamma^2/2\omega_o^2)^{1/2} \equiv \omega_{max} \tag{3.16}$$

follows. The frequency ω_{max} at which the maximum amplitude occurs is a lower frequency than ω_o although we will see that the difference is usually very small. We can find the maximum value of the amplitude A_{max} by substituting ω_{max} in Equation (3.15). The result is

$$A_{max} = \frac{a\omega_o/\gamma}{(1 - \gamma^2/4\omega_o^2)^{1/2}}. \tag{3.17}$$

The dependences of the amplitude $A(\omega)$ and the phase angle δ on the driving frequency ω are shown in Figure 3.5. (We recall that δ is the phase angle by which the displacement lags behind the driving force.) These curves are similar to those for the undamped case (Figure 3.3). With damping, however, the phase angle varies continuously; the maximum amplitude remains finite although large and occurs at a lower frequency than ω_o. Finally, in order to make Equation (3.15) more general, we make use of the substitution $F_0 = ka$, Equation (3.4), to obtain

$$A(\omega) = \frac{F_0/m}{[(\omega_o^2 - \omega^2)^2 + \omega^2\gamma^2]^{1/2}}. \tag{3.18}$$

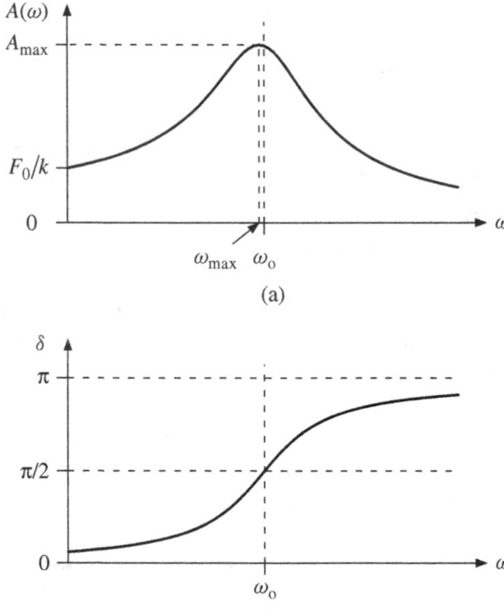

Figure 3.5 (a) A plot of the amplitude $A(\omega)$ of a forced oscillator against the driving frequency ω for the case where damping is present. (b) The variation of the phase angle δ with driving frequency.

The amplitude $A(\omega)$ is proportional to the amplitude F_0 of the applied force and depends on the applied frequency. We emphasise that a periodic driving force can produce oscillations of large amplitude when applied at the resonance frequency. It may then be desirable to add damping to limit the amplitude. For example, sky scrapers will sway in a strong wind. To ensure that the induced oscillations do not reach dangerously high levels, damping mechanisms are included in the construction of such buildings.

In our discussion of damped free oscillations, Section 2.3.1, we defined the quality factor Q of the system by

$$Q = \frac{\omega_o}{\gamma}, \tag{2.21}$$

i.e. as the ratio of the natural frequency ω_o to the damping term γ, essentially a measure of the number of complete oscillations before the oscillations die away. Q also has important significance in the description of forced oscillations as we will see in Section 3.3. In the meantime we use the substitution $Q = \omega_o/\gamma$ in the equations for ω_{max} and A_{max}. Equation (3.16) leads to

$$\omega_{max} = \omega_o(1 - 1/2Q^2)^{1/2}, \tag{3.19}$$

and Equation (3.17) to

$$A_{max} = \frac{aQ}{(1 - 1/4Q^2)^{1/2}}. \tag{3.20}$$

For the case of light damping, when $Q \gg 1$, $\omega_{max} = \omega_o$ and $A_{max} = aQ$ to good approximations. Thus under this condition, the maximum amplitude of oscillation, i.e. resonance, occurs for all practical purposes at the natural frequency of free oscillations ω_o. Moreover, at this frequency we see that the forced oscillator acts like an amplifier with an amplification factor equal to Q.

Worked example

A mass of 1.5 kg rests on a horizontal table and is attached to one end of a spring of spring constant 150 N m^{-1}. The other end of the spring is moved in the horizontal direction according to $x = a\cos\omega t$ where $a = 5 \times 10^{-3}$ m and $\omega = 6\pi$ rad s^{-1}. The damping constant $b = 3.0$ N m^{-1} s. Determine the amplitude and relative phase of the steady state oscillations of the mass. Show that if the applied frequency were adjusted for resonance, the mass would oscillate with an amplitude of approximately 2.5×10^{-2} m.

Solution

$$\omega_o = \sqrt{k/m} = \sqrt{150/1.5} = 10 \text{ rad s}^{-1}, \text{ and } \gamma = b/m = 2.0 \text{ s}^{-1}.$$

$$A(\omega) = \frac{ak/m}{[(\omega_o^2 - \omega^2)^2 + \omega^2\gamma^2]^{1/2}}$$

$$= \frac{150 \times 5 \times 10^{-3}}{1.5\{[10^2 - (6\pi)^2]^2 + 2^2(6\pi)^2\}^{1/2}} = 1.9 \times 10^{-3} \text{ m}.$$

$$\tan\delta = \frac{\omega\gamma}{(\omega_o^2 - \omega^2)} = \frac{6\pi \times 2}{(10^2 - (6\pi)^2)} = -0.15.$$

Since $\omega > \omega_o$, the phase angle must lie between $\pi/2$ and π. Then $\delta = 3.0$ rad. At resonance $\omega \simeq \omega_o$ and $A = A_{max} \simeq a\omega_o/\gamma$, as follows from Equation (3.17), since $\omega_o/\gamma = 5$. Hence $A_{max} \simeq 2.5 \times 10^{-2}$ m.

Worked example

Figure 3.6 shows a schematic diagram of a system that is used to isolate a platform from floor vibrations. The mass of the platform is m, the spring constant of the system is k and there is a damping mechanism (called a dashpot) with damping constant b. If the floor is vibrating according to $\xi = A\cos\omega t$ with respect to its equilibrium position, obtain an expression for the maximum value of the displacement x of the platform from its equilibrium position in terms of A and ω.

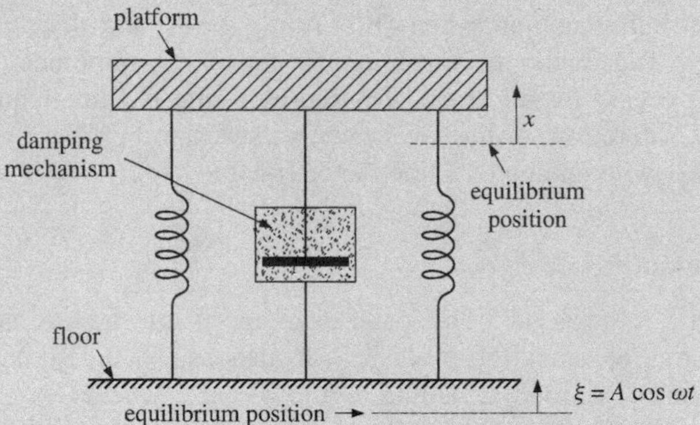

Figure 3.6 Vibration-isolation system showing a platform mounted on springs with a damping mechanism (called a dashpot) with damping factor b.

Solution

The spring force acting on the platform is proportional to the spring extension $(x - \xi)$. The damping force produced by the dashpot is proportional to the relative velocity of the platform with respect to the floor, which is given by

$\frac{\mathrm{d}}{\mathrm{d}t}(x - \xi)$. Thus the equation of motion of the platform is

$$m\frac{\mathrm{d}^2x}{\mathrm{d}t^2} = -k(x - \xi) - b\frac{\mathrm{d}}{\mathrm{d}t}(x - \xi)$$

or

$$\frac{\mathrm{d}^2x}{\mathrm{d}t^2} + \gamma\frac{\mathrm{d}}{\mathrm{d}t}(x - \xi) + \omega_0^2(x - \xi) = 0$$

where $\omega_0^2 = k/m$ and $\gamma = b/m$. To solve this equation we introduce the variable $X = x - \xi$ in place of x, giving the equation

$$\frac{\mathrm{d}^2X}{\mathrm{d}t^2} + \gamma\frac{\mathrm{d}X}{\mathrm{d}t} + \omega_0^2X = -\frac{\mathrm{d}^2y}{\mathrm{d}t^2} = \omega^2A\cos\omega t$$

since $\xi = A\cos\omega t$. We assume a steady state solution of this equation of the form $X = B(\omega)\cos(\omega t - \delta)$ to obtain

$$x = \frac{\omega^2A\cos(\omega t - \delta)}{[(\omega_0^2 - \omega^2)^2 + \omega^2\gamma^2]^{1/2}} + A\cos\omega t.$$

Since x is a superposition of two cosine terms in ωt, we can write it as

$$x = C(\omega)\cos(\omega t - \alpha),$$

where

$$[C(\omega)]^2 = \frac{A^2(\omega_0^4 + \omega^2\gamma^2)}{(\omega_0^2 - \omega^2)^2 + \omega^2\gamma^2}.$$

Thus

$$\left[\frac{C(\omega)}{A}\right]^2 = \frac{(\omega_0^4 + \omega^2\gamma^2)}{(\omega_0^2 - \omega^2)^2 + \omega^2\gamma^2}$$

$$= \frac{(\omega_0^4 + \omega^2\gamma^2)}{(\omega_0^4 + \omega^2\gamma^2) + \omega^2(\omega^2 - 2\omega_0^2)}.$$

The maximum value of x is equal to $C(\omega)$. Inspection of this expression shows that the ratio of $C(\omega)$ to A depends on the relative sizes of the terms $(\omega_0^4 + \omega^2\gamma^2)$ and $\omega^2(\omega^2 - 2\omega_0^2)$:

(i) $C(\omega) = A$ when $\omega^2 = 2\omega_0^2$.
(ii) $C(\omega) > A$ when $\omega^2 < 2\omega_0^2$.
(iii) $C(\omega) < A$ when $\omega^2 > 2\omega_0^2$.

We see that the system does attenuate the floor vibrations when $\omega^2 > 2\omega_0^2$. However, the vibrations of the floor are *amplified* when $\omega^2 < 2\omega_0^2$. It is thus important to make the resonance frequency ω_0 of the system as low as possible by, for example, having a platform of large mass. In practical systems, ω_0 is chosen to be about 1 Hz. Of course, damping can be used to reduce $C(\omega)$ to an acceptable value. For example, at $\omega = \omega_0$, $C(\omega) = A[(\omega_0/\gamma)^2 + 1]^{1/2}$ and increasing the damping factor γ can be seen to reduce $C(\omega)$. Such vibration-isolation systems find many practical applications, such as in tables to support sensitive apparatus like lasers.

3.3 POWER ABSORBED DURING FORCED OSCILLATIONS

In Section 3.2.1 we described the application of a periodic driving force to a mass on the end of a spring, for the ideal situation where there is no damping. The applied force drives the mass back and forth, but if there is no damping there is no dissipation of energy. During steady state oscillations, energy must be provided to stretch or compress the spring but this energy is recovered as the spring returns to its equilibrium length. Consequently, the total power delivered to the oscillator over each complete cycle is zero. However, a real oscillator loses energy because of the frictional damping forces that are invariably present. The driving force has to restore this lost energy. *The power absorbed by the oscillator to sustain its motion is exactly equal to the rate at which the energy is dissipated*. As usual we will assume that the damping force is proportional to the velocity of the mass and so we begin by considering how the velocity varies during forced oscillations. The displacement x of the mass is given by

$$x = A(\omega)\cos(\omega t - \delta) \tag{3.5}$$

where

$$A(\omega) = \frac{a\omega_0^2}{[(\omega_0^2 - \omega^2)^2 + \omega^2\gamma^2]^{1/2}}, \tag{3.15}$$

and so the velocity v is given by

$$v = \frac{\mathrm{d}x}{\mathrm{d}t} = -A(\omega)\omega\sin(\omega t - \delta). \tag{3.21}$$

We write this as

$$v = -v_0(\omega)\sin(\omega t - \delta), \tag{3.22}$$

where

$$v_0(\omega) = A(\omega)\omega. \tag{3.23}$$

We can think of $v_o(\omega)$ as the 'amplitude' of the velocity just as $A(\omega)$ is the amplitude of displacement. Substituting for $A(\omega)$ in Equation (3.23) gives

$$v_o(\omega) = \frac{a\omega_o^2\omega}{[(\omega_o^2 - \omega^2)^2 + \omega^2\gamma^2]^{1/2}}. \tag{3.24}$$

Rewriting $(\omega_o^2 - \omega^2)$ in Equation (3.24) as

$$(\omega_o^2 - \omega^2) = \left(\frac{\omega_o}{\omega} - \frac{\omega}{\omega_o}\right)(\omega_o\omega),$$

we obtain

$$v_o(\omega) = \frac{a\omega_o^2}{\left[\left(\frac{\omega_o}{\omega} - \frac{\omega}{\omega_o}\right)^2 \omega_o^2 + \gamma^2\right]^{1/2}}. \tag{3.25}$$

Inspection of Equation (3.25) shows that

$$\begin{aligned} &\text{as} \quad \omega \to 0, \quad v_o(\omega) \to 0, \\ &\text{as} \quad \omega \to \infty, \quad v_o(\omega) \to 0, \end{aligned}$$

and the value of $v_o(\omega)$ passes through a maximum at exactly $\omega = \omega_o$, when it is equal to $a\omega_o^2/\gamma$.

The rate of energy loss due to damping is equal to the damping force times the velocity of the mass, cf. Equation (2.20). Since the damping force and the velocity are time-dependent, we must define the instantaneous power absorbed at time t by

$$P(t) = bv(t) \times v(t) = b[v(t)]^2.$$

Substituting $v(t)$ from Equation (3.22) gives

$$P(t) = b[v_o(\omega)]^2 \sin^2(\omega t - \delta). \tag{3.26}$$

Furthermore, since the instantaneous power varies it is more appropriate to talk in terms of the average power $\overline{P}(\omega)$ absorbed over a complete cycle of oscillation between times t_o and $t_o + T$, where T is the period. $\overline{P}(\omega)$ is given by

$$\overline{P}(\omega) = \frac{1}{T}\int_{t_o}^{t_o+T} P(t)\mathrm{d}t. \tag{3.27}$$

Thus

$$\overline{P}(\omega) = \frac{b[v_o(\omega)]^2}{T}\int_{t_o}^{t_o+T} \sin^2(\omega t - \delta)\mathrm{d}t. \tag{3.28}$$

The integral of $\sin^2(\omega t - \delta)$ over any complete period of oscillation T is equal to $T/2$. Hence

$$\overline{P}(\omega) = \frac{b[v_o(\omega)]^2}{2}. \tag{3.29}$$

Substituting Equation (3.24) for $v_0(\omega)$ and using $b = m\gamma$, $\omega_o^2 = k/m$ and $a = F_0/k$, we obtain

$$\overline{P}(\omega) = \frac{\omega^2 F_0^2 \gamma}{2m[(\omega_o^2 - \omega^2)^2 + \omega^2\gamma^2]}. \tag{3.30}$$

A plot of $\overline{P}(\omega)$ against ω gives the *power resonance curve* of the oscillator, which shows how the power absorbed by the oscillator varies with the driving frequency. An example of such a power resonance curve is shown in Figure 3.7. Inspection of Equation (3.30) shows that

$$\text{as} \quad \omega \to 0, \quad \overline{P}(\omega) \to 0,$$
$$\text{as} \quad \omega \to \infty, \quad \overline{P}(\omega) \to 0$$

and the maximum value of $\overline{P}(\omega)$ occurs exactly when $\omega = \omega_o$.

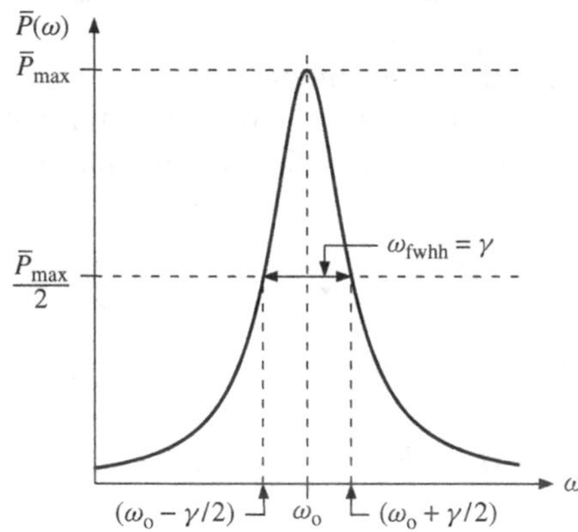

Figure 3.7 The power resonance curve of a forced oscillator. The full width at half height ω_{fwhh} is equal to γ.

An important parameter of a power resonance curve is its full width at half height ω_{fwhh} (see Figure 3.7). This width characterises the *sharpness* of the response of the oscillator to an applied force. When the driving frequency is close to the resonance frequency ω_o, i.e. $\omega \approx \omega_o$, we can replace ω by ω_o everywhere in Equation (3.30) except in the term $(\omega_o^2 - \omega^2)$ which is replaced by

$$(\omega^2 - \omega_o^2) = (\omega_o + \omega)(\omega_o - \omega) \approx 2\omega_o(-\Delta\omega),$$

where $\Delta\omega \equiv \omega - \omega_o$. With these approximations, Equation (3.30) leads to our final expression for the power resonance curve:

$$\boxed{\overline{P}(\omega) = \frac{F_0^2}{2m\gamma(4\Delta\omega^2/\gamma^2 + 1)}.} \tag{3.30a}$$

The maximum value of $\overline{P}(\omega)$ is given by

$$\overline{P}_{\max} = \frac{F_0^2}{2m\gamma} \tag{3.31}$$

and occurs when $\Delta\omega = 0$, i.e. exactly at $\omega = \omega_o$. The half heights of the curve, equal to $\overline{P}_{\max}/2$, occur when $2\Delta\omega/\gamma = 1$, i.e. when $2\Delta\omega = \gamma$. Thus the full width at half height ω_{fwhh} of the resonance curve is given by

$$\omega_{\text{fwhh}} = 2\Delta\omega = \gamma = \omega_o/Q, \tag{3.32a}$$

where the last step follows from the definition (2.21) of the quality factor Q. We see that the full width at half height ω_{fwhh} of the resonance curve is given by the parameter γ. From Equation (3.32a), the quality factor can be written

$$Q = \frac{\omega_o}{\gamma} = \frac{\omega_o}{\omega_{\text{fwhh}}} = \frac{\text{resonance frequency}}{\text{full width at half height of power curve}}. \tag{3.32b}$$

This relationship offers a convenient way to measure the quality factor of an oscillator. Using the relationship $\gamma = \omega_o/Q$, we can rewrite Equation (3.30a) as

$$\overline{P}(\omega) = \frac{F_0^2}{2m\omega_o Q[4(\Delta\omega/\omega_o)^2 + 1/Q^2]}. \tag{3.30b}$$

Power resonance curves for various values of the quality factor Q are presented in Figure 3.8. We see that the higher the value of Q the narrower is the power

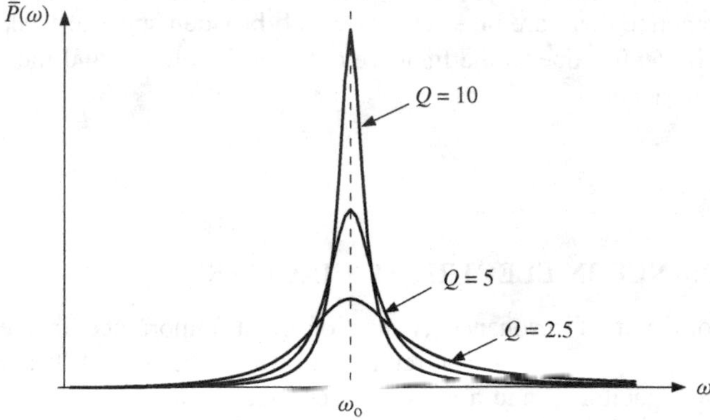

Figure 3.8 Power resonance curves for various values of the quality factor Q.

resonance curve. Moreover, the curves are symmetric about their maxima except for low Q values.

Power resonance curves are common in physical situations. Apart from mechanical and electrical systems, they show up, for example, in atomic and nuclear physics. When an atom is bathed in radiation it may under certain circumstances absorb this radiation. In a classical picture, the oscillating electric field of the radiation interacts with the atom which behaves like a forced oscillator. As for any oscillator of high Q, the atom will only absorb energy over a narrow range of frequencies close to the resonance frequency. This results in a spectral peak in the *absorption spectrum* of the atom where the peak corresponds to a power resonance curve.

Worked example

A spectral peak in the absorption spectrum of an atom occurs at a wavelength of 550 nm and has a measured width of 1.2×10^{-5} nm. Deduce the lifetime of the excited atom.

Solution

$Q = \omega_o/\gamma = \omega_o/\omega_{\text{fwhh}}$, where here ω_{fwhh} is the frequency width of the spectral peak. Then the lifetime of the excited state is given by $1/\gamma = 1/\omega_{\text{fwhh}}$. We are given the width in terms of wavelength λ, where $\omega = 2\pi c/\lambda$. Since $\mathrm{d}\omega = -2\pi c \mathrm{d}\lambda/\lambda^2$,

$$\omega_{\text{fwhh}} \simeq \frac{2\pi c \lambda_{\text{fwhh}}}{\lambda^2}$$

where λ_{fwhh} is the width of the spectral peak in wavelength. Therefore the lifetime of the excited state is equal to

$$\frac{\lambda^2}{2\pi c \lambda_{\text{fwhh}}} = \frac{(550 \times 10^{-9})^2}{2\pi \times 3 \times 10^8 \times 1.2 \times 10^{-14}} = 1.3 \times 10^{-8} \text{ s}.$$

This is the basis of an experimental technique to measure atomic lifetimes. It requires very high photon resolution to determine the widths of the spectral peaks. In practice there are other effects which broaden spectral peaks such as *Doppler broadening* due to the finite velocity of the atoms, and these need to be taken into account.

3.4 RESONANCE IN ELECTRICAL CIRCUITS

The phenomenon of resonance is also of great importance in electrical circuits. An example of a resonance circuit is shown in Figure 3.9. It consists of an inductor L, a capacitor C and a resistor R connected in series, which are driven by an alternative (AC) voltage, $V(t) = V_0 \cos \omega t$. Since there is resistance in the circuit we are dealing with forced oscillations with damping. Applying Kirchoff's law to the circuit gives the equation

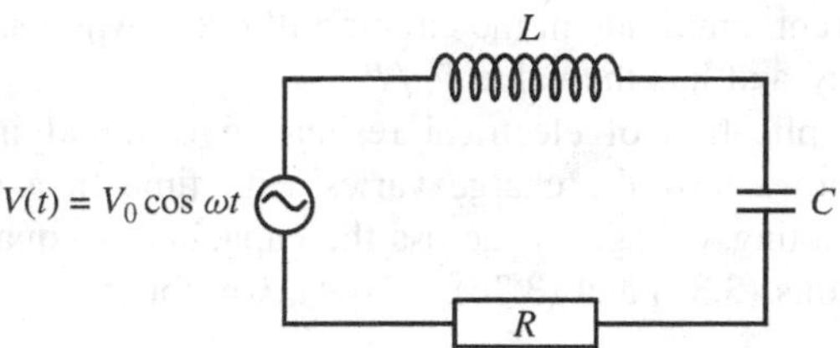

Figure 3.9 An *LCR* resonance circuit that is driven by an alternating voltage $V_0 \cos \omega t$.

$$L\frac{d^2q}{dt^2} + R\frac{dq}{dt} + \frac{q}{C} = V_0 \cos \omega t. \tag{3.33}$$

Comparing this with

$$m\frac{d^2x}{dt^2} + b\frac{dx}{dt} + kx = F_0 \cos \omega t, \tag{3.9}$$

we see that the alternating voltage, $V_0 \cos \omega t$, plays the role of the driving force $F_0 \cos \omega t$, and that m, b and k for the mechanical system are replaced by L, R and $1/C$ for the electrical system. Corresponding replacements in Equations (2.4) and (2.21) give

$$\omega_0^2 = \frac{1}{LC}, \ \gamma = \frac{R}{L}, \ Q = \frac{\omega_0}{\gamma} = \frac{1}{R}\sqrt{\frac{L}{C}}, \tag{3.34}$$

in agreement with our earlier result (2.30). Similarly from the solution, Equations (3.5) and (3.18), of Equation (3.9), it follows that the solution of Equation (3.33) is

$$q = q_0(\omega)\cos(\omega t - \delta) \tag{3.35}$$

where

$$\begin{aligned} q_0(\omega) &= \frac{V_0/L}{[(\omega_0^2 - \omega^2)^2 + (R\omega/L)^2]^{1/2}} \\ &= \frac{V_0}{\omega[(1/\omega C - \omega L)^2 + R^2]^{1/2}} \end{aligned} \tag{3.36}$$

where we have used $\omega_0^2 = 1/LC$. The current I flowing in the circuit is given by

$$\begin{aligned} I = \frac{dq}{dt} &= -q_0(\omega)\omega \sin(\omega t - \delta) \\ &= \frac{V_0 \sin(\omega t - \delta)}{[(1/\omega C - \omega L)^2 + R^2]^{1/2}}. \end{aligned} \tag{3.37}$$

The maximum current amplitude in the circuit will occur when $\omega^2 = \omega_0^2$, i.e. at the resonance frequency and has the value V_0/R.

An important application of electrical resonance is found in radio receivers. Equation (3.36) shows how the charge varies with time in a resonance circuit. The resultant alternating voltage V_C across the capacitor is equal to q/C. Hence, substituting Equations (3.35) and (3.36) for $q(t)$, we obtain

$$V_C = V_C(\omega)\cos(\omega t - \delta),$$

where

$$V_C(\omega) = \frac{V_0/LC}{[(\omega_0^2 - \omega^2)^2 + (R\omega/L)^2]^{1/2}}. \tag{3.38}$$

At resonance when $\omega = \omega_0$, we have

$$V_C(\omega_0) = \frac{V_0}{R\omega_0 C} = QV_0.$$

We see that the resonance circuit has amplified the AC voltage applied to the circuit by the Q-value of the circuit. A typical value of Q might be 200. Moreover, the circuit has been selective in amplifying only those frequencies close to the resonance frequency of the circuit. This makes the circuit ideal for selecting a radio station and amplifying the oscillating radio signal. Figure 3.10 shows a schematic diagram of the input stage of a radio receiver employing an *LCR* resonance circuit. The variable capacitor in Figure 3.10 allows the circuit to be tuned to different radio stations.

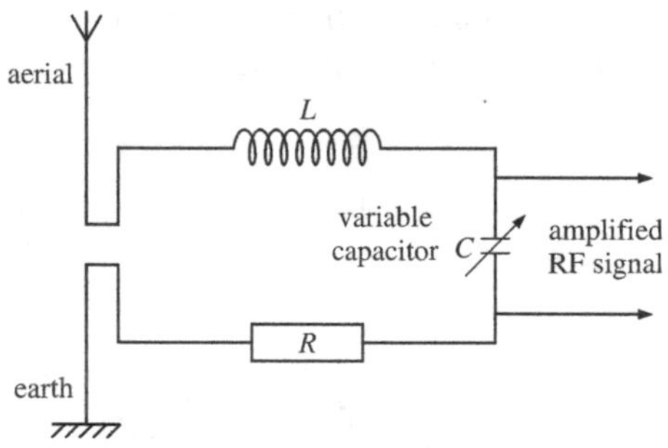

Figure 3.10 A schematic diagram of the input stage of a radio receiver containing an *LCR* resonance circuit. This circuit amplifies the incoming radio signal by a factor equal to the quality factor Q of the circuit. Moreover it amplifies the signal over a narrow range of frequencies which is again determined by the value of Q.

3.5 TRANSIENT PHENOMENA

Our discussion so far has emphasised that the oscillation frequency of a forced oscillator is the same as the frequency ω of the applied driving force. As indicated at the beginning of this chapter, this is not the whole story. When the driving force is first applied and the system is disturbed from its equilibrium position the system will be inclined to oscillate at the frequency of its free oscillations. For the case of light damping, this is essentially the natural frequency ω_o. During this initial period we thus have the sum of two oscillations of frequencies ω and ω_o, respectively. However, as in the case of damped free oscillations (see Section 2.2.1), the oscillations of frequency ω_o die away. The rate at which they do this depends on the degree of damping. The system is then left oscillating at the frequency of the applied force and this is the *steady state* condition. The initial behaviour of the oscillator, before it settles down to the steady state, is referred to as its *transient response*. We can see this mathematically as follows. The equation for damped forced oscillations is

$$\frac{d^2x}{dt^2} + \gamma\frac{dx}{dt} + \omega_o^2 x = \frac{F_0}{m}\cos\omega t. \tag{3.10}$$

If x_1 is a solution of this equation then

$$\frac{d^2x_1}{dt^2} + \gamma\frac{dx_1}{dt} + \omega_o^2 x_1 = \frac{F_0}{m}\cos\omega t.$$

The equation for damped free oscillations is

$$\frac{d^2x}{dt^2} + \gamma\frac{dx}{dt} + \omega_o^2 x = 0. \tag{2.5}$$

If x_2 is a solution of this equation then

$$\frac{d^2x_2}{dt^2} + \gamma\frac{dx_2}{dt} + \omega_o^2 x_2 = 0.$$

Hence

$$\frac{d^2(x_1+x_2)}{dt^2} + \gamma\frac{d(x_1+x_2)}{dt} + \omega_o^2(x_1+x_2) = \frac{F_0}{m}\cos\omega t$$

and so $(x_1 + x_2)$ is also a solution of Equation (3.10). If for x_1 and x_2 in Equations (3.10) and (2.5) we take the solutions given by Equations (3.5) and (2.7), respectively, we obtain as the general solution of Equation (3.10)

$$x = x_1 + x_2 = A(\omega)\cos(\omega t - \delta) + B\exp(-\gamma t/2)\cos[(\omega_o^2 - \gamma^2/4)^{1/2}t + \phi] \tag{3.39}$$

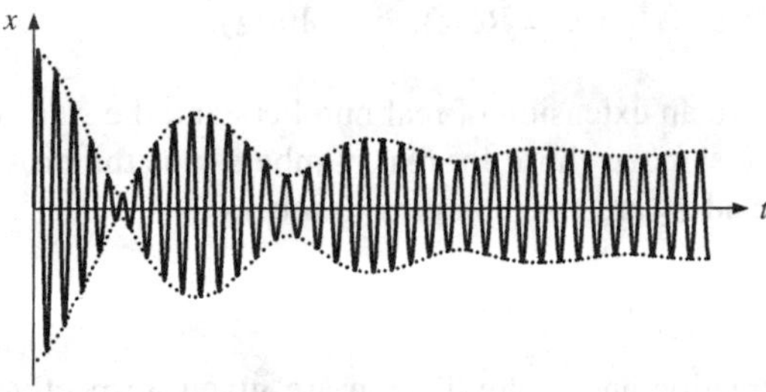

Figure 3.11 An example of the transient response of a forced oscillator. Eventually the oscillations settle down to the steady state condition.

for the case of light damping. The amplitude $A(\omega)$ and the phase angle δ are both functions of driving frequency ω, see Equations (3.13–3.15), and the constants B and ϕ are determined by the initial conditions, as usual. An example of forced oscillations that start at time $t = 0$ is shown in Figure 3.11. After an initial transient response, the system settles down to its steady state condition. Analogous effects occur in AC circuits. When the AC voltage is first applied to the circuit there will be a transient response. This may produce dangerously high voltages and currents, which require special provision in engineering design.

3.6 THE COMPLEX REPRESENTATION OF OSCILLATORY MOTION

Oscillatory motion can also be described using complex numbers. This provides an elegant and concise representation and has important advantages, as we shall see. We start by summarising the relevant mathematical aspects of complex numbers in Section 3.6.1. In Section 3.6.2 we describe how complex numbers are used to represent physical quantities and in Section 3.6.3 we apply complex numbers to the case of forced oscillations with damping.

3.6.1 Complex numbers

A complex number, which is often denoted by z, can be written

$$z = x + iy \tag{3.40}$$

where x and y are real numbers (i.e. ordinary numbers as we have used so far), while i is defined as the square root of -1:

$$i = \sqrt{-1}. \tag{3.41}$$

i is called an *imaginary* number because the square of no real number equals minus one. It follows at once that

$$i^2 = -1. \tag{3.42}$$

We see that a complex number z has two components; a real part x and an imaginary part y, often denoted, respectively, by

$$x = \text{Re}(z), \quad y = \text{Im}(z). \tag{3.43}$$

Complex numbers are an extension of real numbers and the rules of operating with them are exactly the same as those for real numbers plus the proviso that $i^2 = -1$. For example if z_1 and z_2 are two complex numbers

$$z_1 = x_1 + iy_1, \quad z_2 = x_2 + iy_2, \tag{3.44}$$

then addition, subtraction and multiplication are given, respectively, by

$$z_1 \pm z_2 = (x_1 + iy_1) \pm (x_2 + iy_2) = (x_1 \pm x_2) + i(y_1 \pm y_2) \tag{3.45}$$

and

$$z_1 z_2 = (x_1 + iy_1)(x_2 + iy_2) = (x_1 x_2 - y_1 y_2) + i(x_1 y_2 + x_2 y_1). \tag{3.46}$$

The equation $z_1 = z_2$ means $x_1 = x_2$ and $y_1 = y_2$, i.e. the real parts of z_1 and z_2 are equal and so are the imaginary parts. In particular, $z = x + iy = 0$ means that $x = 0$ and $y = 0$. (We can think of the right-hand side of $x + iy = 0$ as standing for $0 + i0$.)

A frequently useful quantity of a complex number is its *complex conjugate*, which is denoted by an asterisk. It is obtained by changing i to $-i$ throughout. Thus the complex conjugate of $z = x + iy$ is

$$z^* = x - iy \tag{3.47}$$

and the complex conjugate of z^2 is

$$(z^2)^* = [(x^2 - y^2) + i2xy]^* = (x^2 - y^2) - i2xy.$$

Using the complex conjugate, it is straightforward to obtain division of complex numbers. To find z_1/z_2 we multiply both the numerator and denominator by the complex conjugate of z_2:

$$\begin{aligned}\frac{z_1}{z_2} = \frac{z_1 z_2^*}{z_2 z_2^*} &= \frac{(x_1 + iy_1)(x_2 - iy_2)}{x_2^2 + y_2^2} \\ &= \frac{(x_1 x_2 + y_1 y_2) + i(x_2 y_1 - x_1 y_2)}{x_2^2 + y_2^2}.\end{aligned} \tag{3.48}$$

The frequently occurring quantity zz^*, i.e. the product of a complex number with its complex conjugate

$$zz^* = x^2 + y^2 \tag{3.49}$$

is seen to be real and positive and is denoted by

$$zz^* = |z|^2. \tag{3.50}$$

The real positive quantity $|z| = \sqrt{zz^*}$ is called the *modulus* of z.

The above summarises the basic rules for manipulating complex numbers. Their meaning is brought out by their geometrical interpretation. We can interpret the components (x, y) of the complex number $z = x + iy$ as the coordinates of a point P in a rectangular Cartesian coordinate system (Figure 3.12). The point P is then specified by the Cartesian coordinates (x, y) or equivalently by the complex number z. The x- and y-axes are called the real and imaginary axes and the whole x-y plane the complex z-plane. Figure 3.12 is referred to as the Argand diagram of z. From Figure 3.12 we see that the distance $OP = \sqrt{(x^2 + y^2)}$ is just the

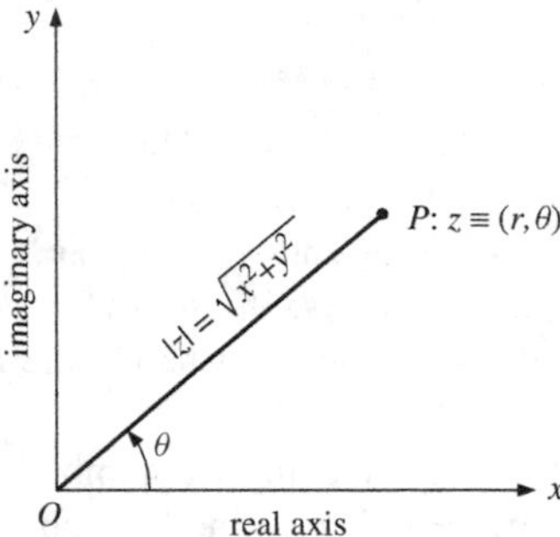

Figure 3.12 The complex plane containing the complex number z.

modulus of z. The angle θ is the angle that the line OP makes with the positive x direction, measured in the anticlockwise sense, and is given by

$$\cos\theta = \frac{x}{\sqrt{(x^2 + y^2)}}, \quad \sin\theta = \frac{y}{\sqrt{(x^2 + y^2)}}. \tag{3.51}$$

Figure 3.12 and these relations suggest the introduction of polar coordinates

$$x = r\cos\theta, \quad y = r\sin\theta \tag{3.52}$$

with r being the distance OP:

$$r = \sqrt{(x^2 + y^2)}. \tag{3.53}$$

The real breakthrough comes through employing the important relation due to Euler[1]

$$\boxed{e^{i\theta} = \cos\theta + i\sin\theta} \tag{3.54}$$

It follows from this relation and Equations (3.40) and (3.51) that

$$z = x + iy = r(\cos\theta + i\sin\theta) = re^{i\theta}. \tag{3.55}$$

The polar coordinate r is the modulus $|z|$ of the complex number z and θ is called the *argument* of z. If we multiply $re^{i\theta}$ by $e^{i\phi}$ we obtain

$$z' = ze^{i\phi} = re^{i(\theta+\phi)}. \tag{3.56}$$

In the Argand diagram, Figure 3.13, this corresponds to rotating the line OP through an angle ϕ in the anticlockwise direction to the new position OP'. If $\phi = \pi/2$ the line is rotated through $\pi/2$. However,

$$e^{i\pi/2} = \cos(\pi/2) + i\sin(\pi/2) = i.$$

[1] A formal verification of this relation is afforded by substituting the power series expansions for $\cos\theta$ and $\sin\theta$ in Equation (3.54). In this way, one obtains the power series expansion of the exponential function $e^{i\theta}$.

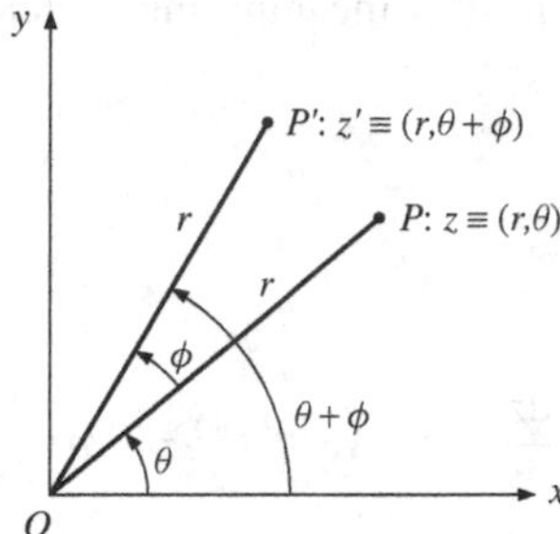

Figure 3.13 Multiplication of the complex number z by the factor $e^{i\phi}$.

Thus multiplying a complex number by $e^{i\pi/2}$ is equivalent to multiplying the number by i. Similarly multiplying a complex number by $e^{i\pi}$ is equivalent to multiplying the number by -1. If θ varies with time as $\theta = (\omega t + \phi)$ then the line OP rotates in the complex plane with angular frequency ω in the anticlockwise direction. As the expression $x = A\cos\theta$ contains both amplitude and angular (or phase) information, so $re^{i\theta}$ also contains these two kinds of information; amplitude information is given by r and phase information is given by θ.

3.6.2 The use of complex numbers to represent physical quantities

The essential idea is that we represent physical quantities such as displacement, velocity and acceleration by the *real part* of a complex number z. We will illustrate this by considering the motion of a simple harmonic oscillator. The complex form of the equation of SHM is

$$\frac{d^2z}{dt^2} = -\omega^2 z, \tag{3.57}$$

where $z = x + iy$. Since x and y are real quantities, taking the real part of this equation, at once gives

$$\frac{d^2x}{dt^2} = -\omega^2 x, \tag{1.6}$$

which is our result from Section 1.2.1. From the solution z of the complex equation (Equation (3.57)) we can take the real part of z to obtain x which is the physically significant quantity.[2] Obtaining a solution of Equation (3.57) in terms of the polar coordinate form $re^{i\theta}$, rather than the equivalent form $z = x + iy$, simplifies the

[2] It should be noted that this procedure of solving a differential equation for a complex variable z, instead of for a real variable x, only works if the equation is linear, i.e. each term in the equation is either independent of z or depends on z or one of its derivatives dz/dt, d^2z/dt^2, ... in first order only. For example, if the right-hand side of Equation (3.57) is replaced by $-\omega^2 z^2$, then $\mathrm{Re}(-\omega^2 z^2) = -\omega^2(x^2 - y^2)$, and taking real parts of the modified equation would lead to $d^2x/dt^2 = -\omega^2(x^2 - y^2)$ and not to $d^2x/dt^2 = -\omega^2 x^2$, the equation we are trying to solve.

analysis and brings out the physical meaning more clearly. Taking for z the polar coordinate form,

$$z = A\mathrm{e}^{i(\omega t+\phi)}, \tag{3.58}$$

then

$$\frac{\mathrm{d}z}{\mathrm{d}t} = i\omega A\mathrm{e}^{i(\omega t+\phi)} = i\omega z \tag{3.59}$$

and

$$\frac{\mathrm{d}^2 z}{\mathrm{d}t^2} = \frac{\mathrm{d}}{\mathrm{d}t}(i\omega z) = -\omega^2 z, \tag{3.60}$$

showing that $z = A\mathrm{e}^{i(\omega t+\phi)}$ is indeed a solution of the SHM equation (3.57). Taking the real parts of Equations (3.58), (3.59) and (3.60) at once gives

$$x = \mathrm{Re}(z) = A\cos(\omega t+\phi),$$
$$\frac{\mathrm{d}x}{\mathrm{d}t} = \mathrm{Re}(i\omega z) = \mathrm{Re}[i\omega(x+iy)] = -\omega y = -\omega A\sin(\omega t+\phi)$$

and

$$\frac{\mathrm{d}^2 x}{\mathrm{d}t^2} = \mathrm{Re}\,(-\omega^2 z) = -\omega^2 A\cos(\omega t+\phi).$$

These are our familiar results for the displacement, velocity and acceleration of a simple harmonic oscillator, cf. Equations (1.11), (1.12) and (1.13).

The geometrical interpretation of complex numbers also provides a representation of physical quantities. Figure 3.14(a) shows $z = Ae^{i(\omega t+\phi)}$ in the complex z-plane. The length of the line OP corresponds to A, the amplitude of the motion, and this line rotates anticlockwise in the complex plane with angular frequency ω. The phase angle ϕ is the angle that the line OP makes with the horizontal axis at time $t = 0$. The projection of OP onto the real axis is equal to $A\cos(\omega t+\phi)$ and corresponds to the physical quantity of displacement x. If we plot this projection as a function of time we obtain the familiar periodic variation of x as shown, for example, in Figure 1.7. Since $i = \mathrm{e}^{i\pi/2}$, Equation (3.59) can be written

$$\frac{\mathrm{d}z}{\mathrm{d}t} = \omega Ae^{i(\omega t+\phi+\pi/2)}. \tag{3.59a}$$

Figure 3.14(b) shows $\mathrm{d}z/\mathrm{d}t$ in the complex plane at point P'. Equation (3.59a) shows that the length of the line OP' is ωA and lies at an angle of $\pi/2$ with respect to the line OP. The physical significance of this is that the velocity in SHM leads the displacement by $\pi/2$, as we saw in Section 1.2.3. The projection of OP' on the real axis is equal to $\omega A\cos(\omega t+\phi+\pi/2)$ and gives the value of the velocity at time t. Writing Equation (3.60) as

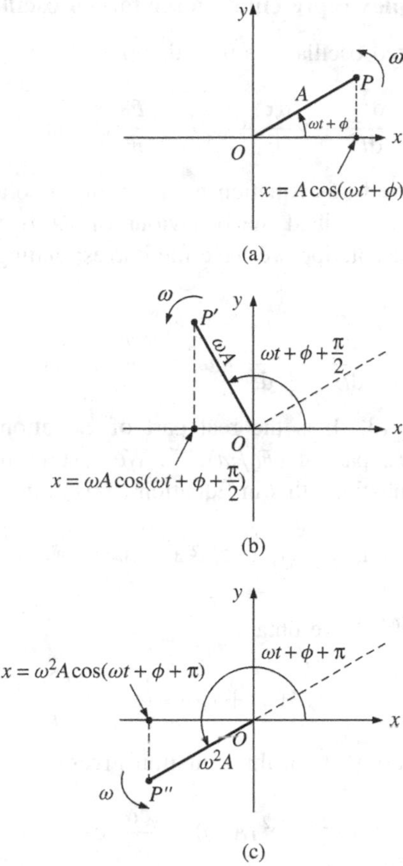

Figure 3.14 Representation of (a) displacement, (b) velocity and (c) acceleration in their respective complex planes. The three lines OP, OP' and OP'' rotate with angular frequency ω maintaining constant phase differences between them.

$$\frac{\mathrm{d}^2 z}{\mathrm{d}t^2} = \omega^2 A e^{i(\omega t+\phi+\pi)}, \tag{3.60a}$$

it follows that the acceleration leads the velocity by $\pi/2$ and leads the displacement by π. The acceleration, given by the projection of OP'' on the real axis, Figure 3.14(c), is equal to $\omega^2 A\cos(\omega t + \phi + \pi)$. The complete picture, then, is of three lines OP, OP' and OP'' rotating anticlockwise at angular frequency ω and maintaining constant relative phases, with their projections on the real axes giving the values of displacement, velocity and acceleration, respectively, as functions of t. This analysis also demonstrates that the mathematical operation of differentiation with respect to time has been replaced by multiplication by $i\omega$. This makes the mathematical manipulation of complex variables much easier than for functions containing sines and cosines.

3.6.3 Use of the complex representation for forced oscillations with damping

The equation for forced oscillations with damping is

$$\frac{d^2x}{dt^2} + \gamma\frac{dx}{dt} + \omega_0^2 x = \frac{F_0}{m}\cos\omega t. \tag{3.10}$$

In Section 3.2.2 we solved this equation by assuming a solution of the form $x = A(\omega)\cos(\omega t - \delta)$ and determined the behaviour of $A(\omega)$ and δ as functions of ω. In the complex representation we have the corresponding complex differential equation

$$\frac{d^2z}{dt^2} + \gamma\frac{dz}{dt} + \omega_0^2 z = \frac{F_0}{m}e^{i\omega t}. \tag{3.61}$$

We note that Equation (3.10) is the real part of Equation (3.61). In particular $(F_0/m)\cos\omega t$ is the real part of $(F_0/m)e^{i\omega t}$. We assume a solution of the form $z = A(\omega)e^{i(\omega t-\delta)}$ and substitute this in Equation (3.61) giving

$$[-\omega^2 A(\omega) + i\gamma\omega A(\omega) + \omega_0^2 A(\omega)]e^{i(\omega t-\delta)} = \frac{F_0}{m}e^{i\omega t}.$$

Dividing through by $e^{i(\omega t-\delta)}$ we obtain

$$(\omega_0^2 - \omega^2)A(\omega) + i\gamma\omega A(\omega) = \frac{F_0}{m}e^{i\delta}. \tag{3.62}$$

Taking real and imaginary parts of this equation gives

$$(\omega_0^2 - \omega^2)A(\omega) = \frac{F_0}{m}\cos\delta$$

and

$$\gamma\omega A(\omega) = \frac{F_0}{m}\sin\delta$$

from which we readily obtain

$$\tan\delta = \frac{\gamma\omega}{(\omega_0^2 - \omega^2)} \tag{3.12}$$

and

$$A(\omega) = \frac{F_0/m}{[(\omega_0^2 - \omega^2)^2 + \omega^2\gamma^2]^{1/2}}. \tag{3.18}$$

These are the same results we obtained in Section 3.2.2 using sines and cosines. However, these results have been obtained more readily using the complex representation.

PROBLEMS 3

3.1 A mass of 0.03 kg rests on a horizontal table and is attached to one end of a spring of spring constant 12 N m^{-1}. The other end of the spring is attached to a rigid support.

The mass is subjected to a harmonic driving force $F = F_0 \cos \omega t$, where $F_0 = 0.15$ N and a damping force $F_d = -bv$, where $b = 0.06$ kg s^{-1}. Determine the amplitude of oscillation and the phase angle between the driving force and the displacement of the mass for steady-state oscillations at frequencies of (a) 2 rad s^{-1}, (b) 20 rad s^{-1} and (c) 100 rad s^{-1}.

3.2 A damped harmonic oscillator, driven by a force $F_0 \cos \omega t$, vibrates with an amplitude $A(\omega)$ given by

$$A(\omega) = \frac{a\omega_o/\omega}{[(\omega_o/\omega - \omega/\omega_o)^2 + 1/Q^2]^{1/2}}$$

where a is the amplitude as $\omega \to 0$, ω_o is the natural frequency of oscillation and Q is the quality factor. Show that the amplitude $A(\omega)$ is a maximum for a frequency

$$\omega_{max} = \omega_o(1 - 1/2Q^2)^{1/2}$$

and that at ω_{max} the amplitude is equal to

$$\frac{aQ}{(1 - 1/4Q^2)^{1/2}}.$$

(Hint: Let $\omega_o/\omega = u$, divide the denominator and numerator by u and investigate the resulting expression inside the square root.)

3.3 For a value of $Q = 10$ in Problem 3.2, find (a) the percentage difference between the natural frequency of oscillation ω_o and the frequency ω_{max} at which the maximum amplitude of oscillation would occur and (b) the percentage difference between the amplitudes at these two frequencies.

3.4 A driven oscillator has a natural frequency ω_o of 100 rad s^{-1}, a Q-value of 25 and an average input power $\overline{P}_{max}$ at resonance of 50 W. Plot the power resonance curve of the oscillator over the frequency range 92 to 108 rad s^{-1}.

3.5 A series *LCR* circuit (cf. Figure 3.9) has $C = 8.0 \times 10^{-6}$ F, $L = 2.0 \times 10^{-2}$ H and $R = 75\ \Omega$ and is driven by a voltage $V(t) = 15 \cos \omega t$ V. Determine (a) the resonance frequency (Hz) of the circuit and (b) the amplitude of the current at this frequency.

3.6 Determine the numerical value of i^i where $i = \sqrt{-1}$.

3.7 The displacement x of a simple harmonic oscillator is given by the real part of the complex number $z = Ae^{i(\omega t + \phi)}$. Derive the phase difference between x and dx/dt, and say which of these is in advance of the other.

3.8 A simple pendulum consists of a mass m attached to a light string of length l. When at rest it lies in a vertical line at $x = 0$. The pendulum is driven by moving its point of suspension harmonically in the *horizontal* direction as $\xi = a \cos \omega t$ about its rest position ($x = 0$). There is a damping force $F_d = -bv$ due to friction as the mass moves through the air with velocity v. (a) Show that the horizontal displacement x of the mass, with respect to its equilibrium position ($x = 0$), is the real part of the complex quantity z where

$$m\frac{d^2z}{dt^2} + b\frac{dz}{dt} + m\omega_o^2 z = m\omega_o^2 a e^{i\omega t}$$

and $\omega_o^2 = g/l$. (b) Assuming a solution of the form $z = Ae^{i(\omega t - \delta)}$, show that the phase angle δ between the driving force and the displacement of the mass is given by

$$\tan \delta = \frac{\gamma\omega}{\omega_o^2 - \omega^2}$$

where $\gamma = b/m$ and that the amplitude is given by

$$A(\omega) = a\frac{\omega_o^2}{[(\omega_o^2 - \omega^2)^2 + \omega^2\gamma^2]^{1/2}}.$$

3.9 When the pendulum in Problem 3.8 is vibrating freely in unforced oscillation, the amplitude of its swing decreases by a factor of e after 75 cycles of oscillation. (a) Determine the Q-value of the pendulum. (b) The point of suspension of the pendulum is moved according to $\xi = a\cos\omega t$ at the resonance frequency ω_o with $a = 0.5$ mm. What will be the amplitude of swing of the pendulum? (c) Show that the width of the *amplitude* resonance curve at half height is equal to $\gamma\sqrt{3}$ and determine its value if the length of the pendulum is 1.5 m. (Assume $g = 9.81\ \mathrm{m\,s^{-2}}$.) (Hint: Follow the approach of Section 3.3 that was used to determine the frequencies at which the half heights of a power resonance curve occur.)

3.10 The equation of motion of a forced harmonic oscillator with damping is given by

$$m\frac{\mathrm{d}^2x}{\mathrm{d}t^2} + b\frac{\mathrm{d}x}{\mathrm{d}t} + kx = F_0\cos\omega t.$$

Assuming a solution $x = A(\omega)\cos(\omega t - \delta)$:

(a) Give expressions for (i) the instantaneous kinetic energy K, (ii) the instantaneous potential energy U and (iii) the instantaneous total energy E of the oscillator.
(b) For what value of ω is the total energy constant with respect to time? What is the total energy of the oscillator at this frequency?
(c) Obtain an expression for the ratio of the average kinetic energy $\overline{K}$ to the average total energy $\overline{E}$ of the oscillator in terms of the dimensionless quantity ω_o/ω. Sketch this expression over an appropriate range of ω. For what value of ω are the average values of the kinetic and potential energies equal?
(d) Show that the average total energy of the oscillator varies with angular frequency ω according to

$$\overline{E}(\omega) = \frac{F_o^2(\omega_o^2 + \omega^2)}{4m[(\omega_o^2 - \omega^2)^2 + \omega^2 b^2/m^2]}.$$

3.11 (a) For a driven oscillator show that the energy dissipated per cycle by a frictional force $F_d = -bv$ at frequency ω and amplitude A is equal to $\pi b\omega A^2$.
(b) Hence show

$$\frac{\text{energy dissipated/cycle}}{\text{stored energy}} = \frac{2\pi b}{m\omega}.$$

(c) Show that at the resonance frequency of a lightly damped oscillator

$$\frac{\text{energy dissipated/cycle}}{\text{stored energy}} = \frac{2\pi}{Q}$$

where Q is the quality factor.

3.12 The pendulum of a clock consists of a mass of 0.20 kg hanging from a thin rod. The amplitude of the pendulum swing is 3.0 cm. The clock is driven by a weight of mass 4.5 kg that falls a distance of 0.95 m over a period of 8 days. Assuming the pendulum to be a simple pendulum of length 0.75 m, show that the Q-value of the clock is approximately 70.
(Assume $g = 9.81\ \mathrm{m\ s^{-2}}$.)

4

Coupled Oscillators

So far we have considered simple harmonic oscillators such as a mass on a spring or a simple pendulum that have only one way of oscillating. These are characterised by a single natural frequency of oscillation. In this chapter we consider systems that consist of two (or more) oscillators that are coupled together in some way and that have more than one frequency of oscillation. We will see that this coupling produces new and important physical effects. Each of the frequencies relates to a different way in which the system can oscillate. These different ways are called *normal modes* and the associated frequencies are called *normal frequencies*. The normal modes of a system are characterised by the fact that *all* parts of the system oscillate with the *same* frequency. Coupled motion is important because oscillators rarely exist in complete isolation and real physical systems are usually capable of oscillating in many different ways. For example a noisy old car will have many coupled components that may be heard vibrating and rattling when the engine is running! At the microscopic level, vibrating atoms in a crystal provide an example of coupled oscillators. Coupled oscillators are also important because they pave the way to the understanding of waves in continuous media like taut strings. Wave motion depends on neighbouring vibrating systems that are coupled together and so can transmit their energy from one to another.

4.1 PHYSICAL CHARACTERISTICS OF COUPLED OSCILLATORS

We can see the main physical characteristics of coupled oscillators by observing the motion of two simple pendulums that are coupled together. They can be coupled by attaching their points of suspension to a supporting string as shown in Figure 4.1. This is a simple experiment that is well worth doing. Both pendulums have the same length l and so their periods of oscillation are equal. The supporting string provides the coupling between the two pendulums. As each pendulum oscillates it pulls on the supporting string and causes the point of suspension of the other

Vibrations and Waves George C. King

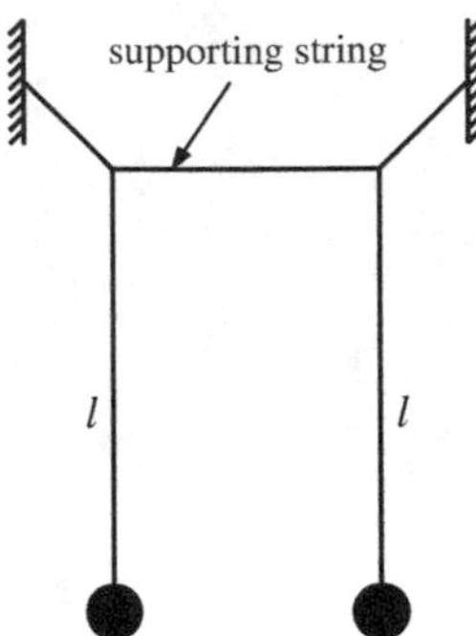

Figure 4.1 Two simple pendulums of length l coupled together by a supporting string. The displacements of the two pendulum masses are considered in the direction perpendicular to the plane of the page.

pendulum to be driven back and forth. The motion of each pendulum affects the other and so their motions cannot be considered in isolation. We consider the motion of the two pendulums in the direction at right angles to the plane of the page. (i) First we displace both pendulum masses by the same amount and in the same direction. When released we observe that the two masses move back and forth in the same directions as each other with the same frequency and the same amplitude. (In this example and for the rest of this chapter we will assume that damping forces can be neglected.) (ii) Next we displace the two masses by the same amount but now in opposite directions. When released the two masses move back and forth in opposite directions. Again they both oscillate with the same frequency as each other but at a frequency that is slightly different from when they move in the same directions. These two distinctly different ways of oscillation are the normal modes of the system. We observe that once the system is put into one or other of these normal modes it stays in that mode and does not evolve into the other. (iii) Now we displace just one mass leaving the other at its equilibrium position. When released the displaced mass moves back and forth but it does so with a steadily decreasing amplitude. At the same time the mass that was initially at rest starts to oscillate and gradually the amplitude of its oscillation increases. Eventually the first mass momentarily stops oscillating having transferred all of its energy to the second mass that now oscillates with the amplitude initially given to the first mass. This process then repeats with the amplitude of the second mass steadily decreasing and that of the first steadily increasing. The cycle continues with the energy repeatedly being transferred between the two masses. This behaviour seems to be strange at first sight and indeed is sometimes used by conjurors to mystify their audience; they might use coconuts as the pendulum masses! However, there is nothing mysterious about the observations. What we are observing is the *superposition* of the two normal modes described above, as we shall see.

4.2 NORMAL MODES OF OSCILLATION

To obtain a mathematical description of coupled oscillations we start again with a pair of simple pendulums but now the coupling is provided by a light horizontal spring that connects them, as shown in Figure 4.2. The spring is at its unstretched length when the two pendulums are at their equilibrium positions. The mass and length of each pendulum are m and l, respectively, and the spring constant is k. Displacements of the two masses from their equilibrium positions are x_a and x_b, respectively, and now, in contrast to Section 4.1, we consider oscillations in the plane of the page.

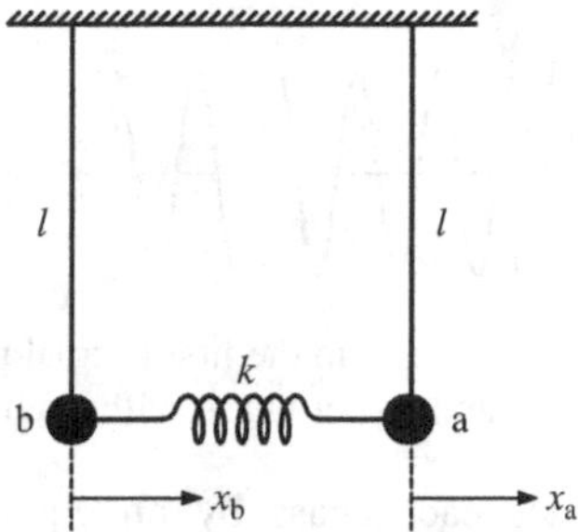

Figure 4.2 Two simple pendulums coupled together by a light horizontal spring of spring constant k. The displacements of the two pendulum masses from their equilibrium positions are x_a and x_b, respectively, and these lie in the plane of the page.

Case (i). We first displace each mass in the same direction by an equal amount as shown in Figure 4.3 and then release them. Since the pendulums have the same period the spring retains its unstretched length and so plays no role in the motion. The two pendulums might just as well be unconnected as they both oscillate at the frequency of a simple pendulum $\sqrt{g/l}$. We can then write the displacements of the two masses, respectively, as

$$x_a = A\cos\omega_1 t, \qquad x_b = A\cos\omega_1 t \tag{4.1}$$

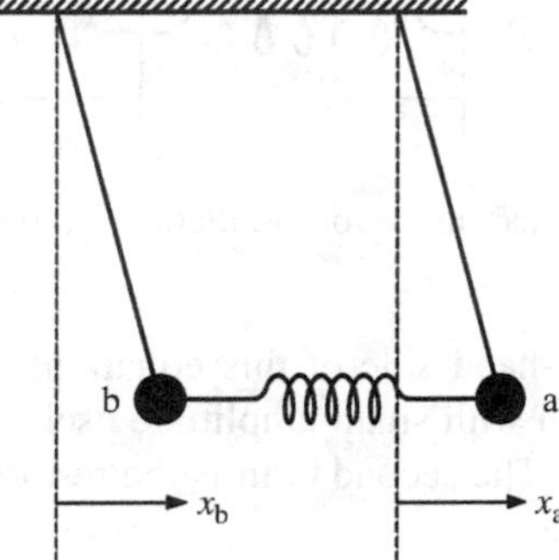

Figure 4.3 The first normal mode of oscillation of the coupled system in which $x_a = x_b$.

where A is the initial displacement and $\omega_1 = \sqrt{g/l}$. The phase angles are zero because the masses start from rest (cf. Section 1.2.4). The variations of x_a and x_b

with time are shown in Figure 4.4. The masses oscillate *in phase* with the same frequency and amplitude. This is the first normal mode of oscillation.

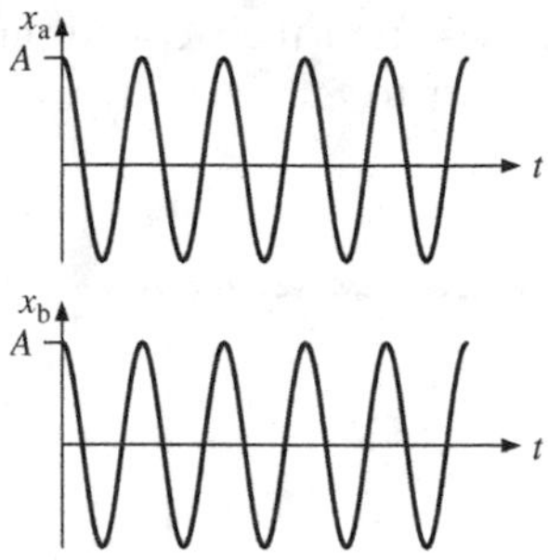

Figure 4.4 Oscillations of the two masses in the first normal mode. These oscillations have the same frequency and amplitude and are in phase with each other.

Case (ii). We now displace each mass by an equal amount but in opposite directions, as shown in Figure 4.5, and then release them. As the two pendulums swing back and forth the spring is alternately stretched and compressed and this exerts an additional restoring force on the masses. The symmetry of the arrangement tells us that the motions of the masses will be mirror images of each other, i.e. $x_a = -x_b$. The resultant equation of motion of mass a is then

$$m\frac{d^2x_a}{d^2t} = -\frac{mgx_a}{l} - 2kx_a. \tag{4.2}$$

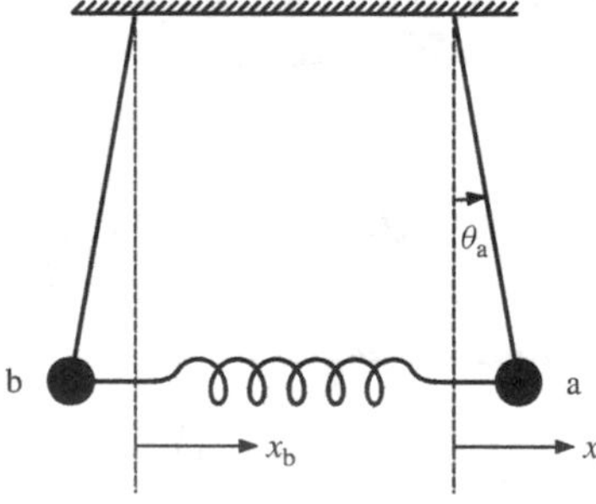

Figure 4.5 The second normal mode of oscillation of the coupled system in which $x_a = -x_b$.

The first term on the right-hand side of this equation is the usual restoring force term for a simple pendulum with small amplitude oscillations [see Equation (1.31) with $x_a \simeq l\theta_a$ for small θ_a]. The second term is the restoring force due to the spring extension of $2x_a$. Hence

$$\frac{d^2x_a}{d^2t} + \omega_2^2 x_a = 0 \tag{4.3}$$

where $\omega_2^2 = (g/l + 2k/m)$. The action of the spring is to increase the restoring force acting on each mass and this increases the frequency of oscillation, i.e. $\omega_2 > \omega_1$.

The solution of Equation (4.3) is

$$x_a = B\cos\omega_2 t, \tag{4.4}$$

where B is the initial displacement. Again the phase angle is zero because the mass started from rest. Since $x_a = -x_b$,

$$x_b = -B\cos\omega_2 t. \tag{4.5}$$

The variations of x_a and x_b with time are shown in Figure 4.6. The masses oscillate with the same frequency and amplitude but now they are 180° *out of phase*. We could write x_b as $x_b = B\cos(\omega_2 t + \pi)$ to emphasise this phase relationship. This is the second normal mode of oscillation. We see that in each normal mode:

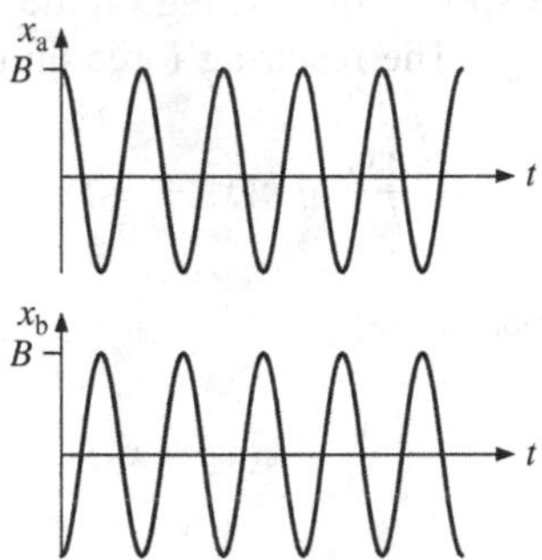

Figure 4.6 Oscillations of the two masses in the second normal mode. These oscillations have the same frequency and amplitude but are in anti-phase, i.e. are 180° out of phase with each other.

- Both the masses oscillate at the *same* frequency.
- Each of the masses performs SHM with constant amplitude.
- There is a well defined phase difference between the two masses; either zero or π.
- Once started in a particular normal mode, the system stays in that mode and does not evolve into the other one.

The importance of normal modes, as we shall see, is that they are entirely independent of each other.

4.3 SUPERPOSITION OF NORMAL MODES

In general the motion of a coupled oscillator will be much more complicated than in cases (i) and (ii) above. Those cases were special in that the motion was confined to a single normal mode, i.e. either $x_a = x_b$ or $x_a = -x_b$ at all times. In general this is not so. The general case is illustrated in Figure 4.7 which shows the displacements of the two masses at some instant in time and $x_a \neq \pm x_b$. This gives a spring extension $(x_a - x_b)$ and produces a tension $T = k(x_a - x_b)$ in the

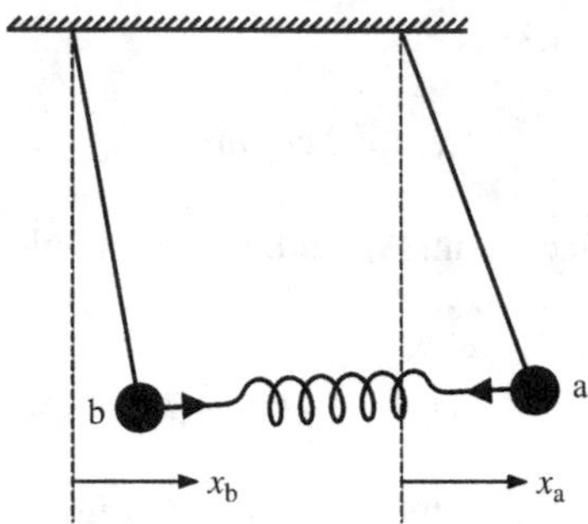

Figure 4.7 General case for the superposition of normal modes where $x_a \neq \pm x_b$.

spring. The directions of the spring force acting on the masses are as indicated by the arrow heads on the springs. The restoring force on mass a is

$$-\frac{mgx_a}{l} - k(x_a - x_b)$$

and the restoring force on mass b is

$$-\frac{mgx_b}{l} + k(x_a - x_b).$$

The resultant equations of motion are

$$\frac{d^2x_a}{dt^2} + \frac{gx_a}{l} + \frac{k}{m}(x_a - x_b) = 0 \quad (4.6)$$

and

$$\frac{d^2x_b}{dt^2} + \frac{gx_b}{l} - \frac{k}{m}(x_a - x_b) = 0. \quad (4.7)$$

Equations (4.6) and (4.7) each contain both x_a and x_b. Thus they cannot be solved separately but must be solved simultaneously. We can do this as follows. Adding them gives

$$\frac{d^2(x_a + x_b)}{dt^2} + \frac{g(x_a + x_b)}{l} = 0. \quad (4.8)$$

It is striking that this is the equation of SHM where the variable is $(x_a + x_b)$. Moreover the frequency of oscillation $\sqrt{g/l}$ is the frequency ω_1 of the first normal mode. Subtracting Equation (4.7) from Equation (4.6) gives

$$\frac{d^2(x_a - x_b)}{dt^2} + \left(\frac{g}{l} + \frac{2k}{m}\right)(x_a - x_b) = 0. \quad (4.9)$$

This again is the equation of SHM but now in the variable $(x_a - x_b)$. Moreover, the oscillation frequency $\sqrt{(g/l + 2k/m)}$ is the same as the frequency ω_2 of the second normal mode. We introduce the new variables q_1 and q_2 where

$$q_1 = (x_a + x_b) \text{ and } q_2 = (x_a - x_b). \tag{4.10}$$

Then

$$\frac{d^2 q_1}{dt^2} + \omega_1^2 q_1 = 0 \tag{4.11}$$

and

$$\frac{d^2 q_2}{dt^2} + \omega_2^2 q_2 = 0. \tag{4.12}$$

We now have another description of the normal modes. We have two *independent* oscillations in which each normal mode is represented by the oscillation of a single variable: each of Equations (4.11) and (4.12) involves just one coordinate, q_1 or q_2, and describes SHM, with frequencies ω_1 and ω_2, respectively. These equations do not involve, for example, a product $q_1 q_2$: there is no coupling between the two normal modes. This is in contrast to Equations (4.6) and (4.7) which contain both position coordinates x_a and x_b. The terms in those equations involving $(x_a - x_b)$ represent the effect that each mass has on the other via the connecting spring. They couple the oscillations of the two masses: the oscillations are *not* independent. The general solutions of Equations (4.11) and (4.12) can be written, respectively,

$$q_1 = C_1 \cos(\omega_1 t + \phi_1), \qquad q_2 = C_2 \cos(\omega_2 t + \phi_2), \tag{4.13}$$

as we know from Section 1.2.4. C_1 and C_2 are amplitudes and ϕ_1 and ϕ_2 are phase angles. The variables q_1 and q_2 are called *normal coordinates* and ω_1 and ω_2 are called *normal frequencies*. If $q_1 = 0$ then $x_a = -x_b$ at all times, and if $q_2 = 0$ then $x_a = x_b$ at all times. It is useful to describe coupled motion in terms of the normal coordinates because the resulting equations of motion depend on *only one* variable, either q_1 or q_2, so that they can be considered separately; changes in q_1 do not affect q_2 and vice versa. For example, the amplitude and hence energy of each normal mode remains constant; energy *never* flows between one normal mode and another as will be demonstrated shortly.

We can express the displacements of the two masses in terms of the normal coordinates. Equation (4.10) leads to

$$x_a = \frac{1}{2}(q_1 + q_2) = \frac{1}{2}[C_1 \cos(\omega_1 t + \phi_1) + C_2 \cos(\omega_2 t + \phi_2)] \tag{4.14}$$

and

$$x_b = \frac{1}{2}(q_1 - q_2) = \frac{1}{2}[C_1 \cos(\omega_1 t + \phi_1) - C_2 \cos(\omega_2 t + \phi_2)]. \tag{4.15}$$

We see that the apparently complicated motion of a coupled oscillator (see Section 4.1) can be broken down into a combination of two independent harmonic oscillations (normal modes). The variables of these harmonic motions are the normal coordinates. Equations (4.14) and (4.15) demonstrate that *any* solution of Equations (4.6) and (4.7), i.e. *any* motion of the two masses, can be written as a *superposition* of the two normal modes. It follows that there are just two normal modes for our system. The four constants C_1, C_2, ϕ_1 and ϕ_2 are determined by the initial positions and velocities of the two masses, i.e. at time $t = 0$. If the two masses are released from rest at $t = 0$, the appropriate solutions for q_1 and q_2, obtained by taking $\phi_1 = \phi_2 = 0$ in Equation (4.13), are

$$q_1 = C_1 \cos \omega_1 t \text{ and } q_2 = C_2 \cos \omega_2 t. \tag{4.16}$$

The independence of the two normal modes is clearly demonstrated if we write down the energy of the system. In terms of the position coordinates x_a and x_b the energy is given by

$$E = \frac{1}{2}m\left(\frac{dx_a}{dt}\right)^2 + \frac{1}{2}m\left(\frac{dx_b}{dt}\right)^2 + \frac{1}{2}\frac{mg}{l}(x_a^2 + x_b^2) + \frac{1}{2}k(x_a - x_b)^2. \tag{4.17a}$$

The first two terms in this expression are the kinetic energies of the two masses, the third term is their potential energies due to gravity [see Equation (1.36)] and the last term is the energy stored in the spring [see Equation (1.18)]. Expressed in terms of the normal coordinates q_1 and q_2 (Equation (4.10)) the energy E becomes

$$E = \left[\frac{1}{4}m\left(\frac{dq_1}{dt}\right)^2 + \frac{1}{4}\left(\frac{mg}{l}\right)q_1^2\right] + \left[\frac{1}{4}m\left(\frac{dq_2}{dt}\right)^2 + \frac{1}{4}\left(\frac{mg}{l} + 2k\right)q_2^2\right]. \tag{4.17b}$$

This equation represents the energy of two *independent* simple harmonic oscillators with frequencies $\omega_1 = \sqrt{g/l}$ and $\omega_2 = \sqrt{(g/l + 2k/m)}$ (see also the discussion in Section 1.3.2). Each of the expressions in square brackets in this equation contains only one of the normal coordinates and represents the energy of a single isolated harmonic oscillator. There are no 'cross terms' involving both q_1 and q_2, which would indicate coupling between them. This is in contrast to the energy expressed in terms of the position coordinates x_a and x_b (Equation (4.17a)) where the last term, involving $(x_a - x_b)$, represents a coupling between the two masses.

Worked example

Consider the system of two identical simple pendulums connected by a light horizontal spring. Deduce expressions for the displacement of the two masses in terms of the normal modes of the system for the following sets of initial conditions, (at $t = 0$). In all cases the masses are released from rest. (i) $x_a = A, x_b = A$, (ii) $x_a = A, x_b = -A$ and (iii) $x_a = A, x_b = 0$.

Solution

We have $x_a = \frac{1}{2}(C_1 \cos\omega_1 t + C_2 \cos\omega_2 t)$ and $x_b = \frac{1}{2}(C_1 \cos\omega_1 t - C_2 \cos\omega_2 t)$.

(i) Substituting for $x_a = A, x_b = A$ at $t = 0$ gives

$$A = \frac{1}{2}(C_1 + C_2) \text{ and } A = \frac{1}{2}(C_1 - C_2).$$

Hence $C_1 = 2A$ and $C_2 = 0$, giving $x_a = A\cos\omega_1 t$ and $x_b = A\cos\omega_1 t$. We recognise this as the first normal mode with *all* the motion in this mode with frequency ω_1.

(ii) Substituting for $x_a = A, x_b = -A$ at $t = 0$ gives

$$A = \frac{1}{2}(C_1 + C_2) \text{ and } -A = \frac{1}{2}(C_1 - C_2).$$

Hence $C_1 = 0$ and $C_2 = 2A$, giving $x_a = A\cos\omega_2 t$ and $x_b = -A\cos\omega_2 t$. We recognise this as the second normal mode with *all* the motion in this mode with frequency ω_2.

(iii) Substituting for $x_a = A, x_b = 0$ at $t = 0$ gives

$$A = \frac{1}{2}(C_1 + C_2) \text{ and } 0 = \frac{1}{2}(C_1 - C_2).$$

Hence $C_1 = A$ and $C_2 = A$, giving

$$x_a = \frac{1}{2}(A\cos\omega_1 t + A\cos\omega_2 t) \text{ and } x_b = \frac{1}{2}(A\cos\omega_1 t - A\cos\omega_2 t).$$

These equations for x_a and x_b combine equal amounts of the two normal modes. We can visualise these results in a different way by recasting the solutions for x_a and x_b as follows. Recalling the trigonometrical identities:

$$\cos(\alpha \pm \beta) = \cos\alpha\cos\beta \mp \sin\alpha\sin\beta,$$

we obtain

$$\cos(\alpha - \beta) + \cos(\alpha + \beta) = 2\cos\beta\cos\alpha.$$

Letting $(\alpha - \beta) = \omega_1$ and $(\alpha + \beta) = \omega_2$ we obtain

$$\alpha = \frac{(\omega_2 + \omega_1)}{2} \text{ and } \beta = \frac{(\omega_2 - \omega_1)}{2}.$$

Thus

$$\cos\omega_1 t + \cos\omega_2 t = 2\cos\frac{(\omega_2 - \omega_1)t}{2}\cos\frac{(\omega_2 + \omega_1)t}{2}$$

giving

$$x_a = A\cos\frac{(\omega_2-\omega_1)t}{2}\cos\frac{(\omega_2+\omega_1)t}{2}.$$

This product represents a high frequency oscillation at the mean of the two normal frequencies whose amplitude is modulated by a low frequency term at half the difference in frequency. This is completely analogous to the phenomena of beating that occurs when two sound waves of slightly different frequency combine (see also Section 8.1.1). The beats that we hear arise from the low frequency modulation term. In a similar way we find

$$x_b = A\sin\frac{(\omega_2-\omega_1)t}{2}\sin\frac{(\omega_2+\omega_1)t}{2},$$

which we can write as

$$x_b = A\cos\left[\frac{(\omega_2-\omega_1)t}{2}-\frac{\pi}{2}\right]\cos\left[\frac{(\omega_2+\omega_1)t}{2}-\frac{\pi}{2}\right].$$

Again we have a high frequency oscillation modulated by a low frequency term. We see, however, that both cosine terms in the expression for x_b are exactly $\pi/2$ out of phase with respect to the corresponding terms for x_a. The variations of x_a and x_b with time are plotted in Figure 4.8. These results explain the behaviour of the two coupled pendulums in Section 4.1, where one pendulum was given an initial displacement and the other was initially at its equilibrium position. The important point in all of these examples, with different initial conditions, is that the subsequent motion is always a superposition of the normal modes.

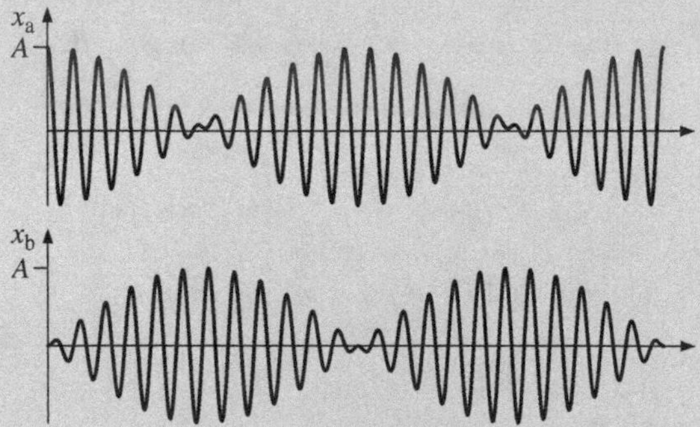

Figure 4.8 Oscillations of the coupled pendulums which, occur when one mass was initially ($t = 0$) at $x_a = A$ and the other at $x_b = 0$.

4.4 OSCILLATING MASSES COUPLED BY SPRINGS

We now consider the case of oscillating masses coupled together by springs. Figure 4.9 shows two identical but independent mass-spring oscillators with mass m and spring constant k attached to two rigid walls, cf. Figure 1.1. The two oscillators are coupled together by a third spring also of spring constant k as shown in Figure 4.10. This third spring provides the coupling so that the motion of one mass influences the motion of the other. This coupled system has two normal modes of oscillation. We wish to determine the two frequencies at which the system will oscillate, i.e. the normal frequencies and the relative displacements of the masses in the two normal modes. We could exploit the symmetry of the system to spot the two normal modes as we did in Section 4.2 for the coupled pendulums. Our physical intuition would suggest that the normal modes would be (i) where both masses move in the same direction and (ii) where they move in opposite directions. These two modes are indicated by the arrows in Figure 4.10. We might also expect that mode (ii) would have the higher frequency of oscillation since all three springs are having an effect rather than just two as in mode (i). Instead of spotting the normal modes we adopt a more general approach where we make use of the characteristics of normal modes, namely that in a normal mode all of the masses oscillate at the same frequency and each mass performs SHM with constant amplitude. For the sake of simplicity we will assume that the two masses are initially at rest, i.e. they have zero velocity at $t = 0$. Figure 4.11 shows

Figure 4.9 Two uncoupled mass-spring oscillators.

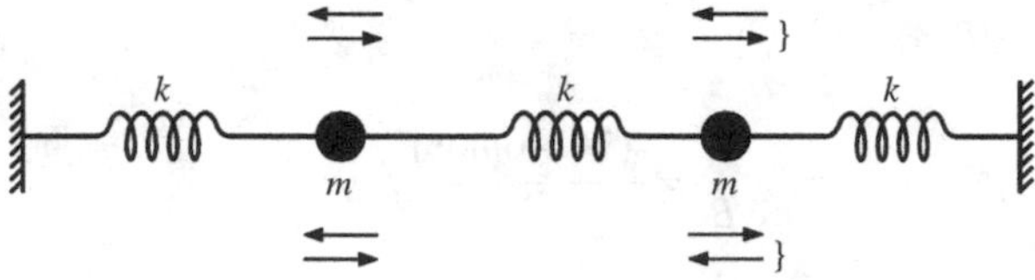

Figure 4.10 Two mass-spring oscillators coupled together by a third spring. The arrows indicate the directions of the displacements of the two masses expected in the two normal modes.

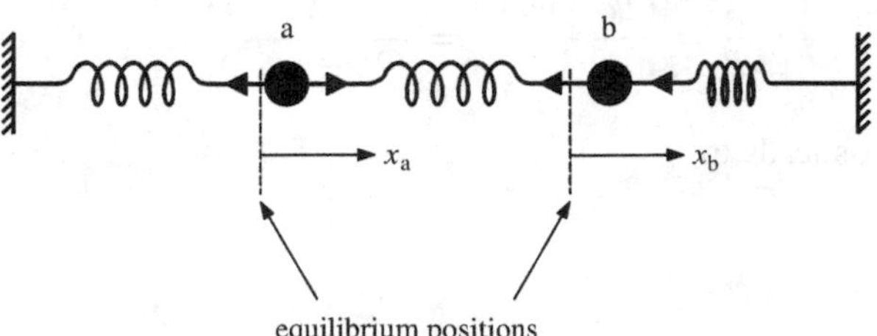

Figure 4.11 Two mass-spring oscillators coupled together by a third spring. The masses are at arbitrary displacements, x_a and x_b, respectively, from their equilibrium positions.

the two masses a and b displaced by arbitrary values x_a and x_b, respectively, from their equilibrium positions at some instant in time. In order to see more easily the directions of the forces acting on the masses we let $x_b > x_a$. The left-hand spring is extended by x_a, the middle spring is stretched by $(x_b - x_a)$ and the right-hand spring is compressed by x_b. The directions of the resultant forces on the masses are shown by the directions of the arrow heads. To obtain the equation of motion for each mass we need to consider only the forces exerted by the springs on either side of the mass. The resultant equations of motion are

$$m\frac{d^2x_a}{dt^2} = -kx_a + k(x_b - x_a) = kx_b - 2kx_a \tag{4.18}$$

and

$$m\frac{d^2x_b}{dt^2} = -k(x_b - x_a) - kx_b = kx_a - 2kx_b. \tag{4.19}$$

We are looking for normal mode solutions of these equations, where both masses oscillate at the same frequency ω, i.e. solutions of the form $x_a = A\cos\omega t$ and $x_b = B\cos\omega t$. Substituting for x_a in Equation (4.18) yields

$$-Am\omega^2\cos\omega t = kB\cos\omega t - 2kA\cos\omega t,$$

giving

$$\frac{A}{B} = \frac{k}{(2k - m\omega^2)}. \tag{4.20}$$

Substituting for x_b in Equation (4.19) yields

$$-Bm\omega^2\cos\omega t = kA\cos\omega t - 2kB\cos\omega t,$$

giving

$$\frac{A}{B} = \frac{(2k - m\omega^2)}{k}. \tag{4.21}$$

So long as A and B are not both zero, the right-hand sides of Equations (4.21) and (4.22) must be equal, i.e. we require

$$\frac{A}{B} = \frac{(2k - m\omega^2)}{k} = \frac{k}{(2k - m\omega^2)}. \tag{4.22}$$

Multiplying across leads to

$$(2k - m\omega^2)^2 = k^2. \tag{4.23}$$

This is a quadratic equation in ω^2 which is seen at once to have the solutions $(2k - m\omega^2) = \pm k$, i.e. $\omega^2 = k/m$ or $3k/m$. These are the two normal frequencies of the coupled system. Putting $\omega^2 = k/m$ in Equation (4.20) gives $A = B$. This is the first normal mode in which the two masses move in the same direction as each other and with the same amplitude. Then

$$x_a = A\cos\omega_1 t, \qquad x_b = A\cos\omega_1 t, \tag{4.24}$$

where $\omega_1^2 = k/m$. Putting $\omega^2 = 3k/m$ in Equation (4.20) gives $A = -B$. This is the second normal mode where the minus sign tells us that the masses move in opposite directions. Thus

$$x_a = A\cos\omega_2 t, \qquad x_b = -A\cos\omega_2 t, \tag{4.25}$$

where $\omega_2^2 = 3k/m$. All of these results are in agreement with our physical intuition. Since most coupled oscillators do not have a symmetry that allows us to spot the normal modes, the approach described here is normally essential. As usual the general motion will be a superposition of the two normal modes, i.e.

$$x_a = C_1\cos\omega_1 t + C_2\cos\omega_2 t$$

and

$$x_b = C_1\cos\omega_1 t - C_2\cos\omega_2 t.$$

If the masses did not have zero velocity at $t = 0$, we would also need to include phase angles as in Equations (4.14) and (4.15).

Worked example

Figure 4.12 shows two equal masses of mass m suspended from two identical springs of spring constant k. Determine the normal frequencies of this system

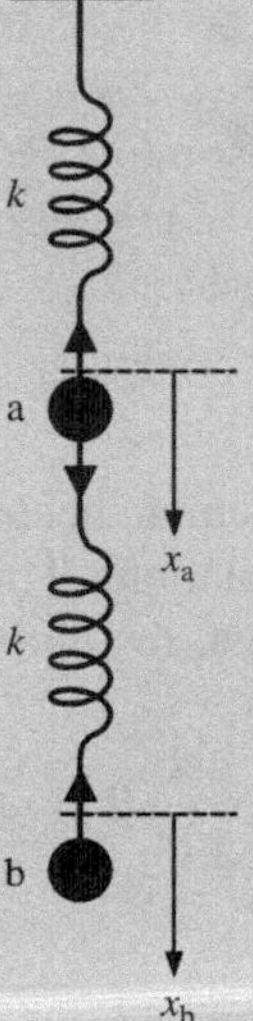

Figure 4.12 Two equal masses m suspended from two identical springs of spring constant k. The displacements of the two masses from their equilibrium positions are x_a and x_b respectively, measured in the downward direction.

for vertical oscillations and the ratios of the amplitudes of oscillation of the masses at these frequencies.

Solution

Let x_a and x_b be arbitrary displacements of the masses from their respective equilibrium positions and let x_b be greater than x_a. Then the extensions of the upper and lower springs are x_a and $(x_b - x_a)$, respectively, and the directions of the forces acting upon the two masses are as indicated by the arrow heads. The resultant equations of motion are

$$m\frac{d^2x_a}{dt^2} = -kx_a + k(x_b - x_a) = k(x_b - 2x_a)$$

and

$$m\frac{d^2x_b}{dt^2} = -k(x_b - x_a).$$

This time we try complex solutions of the form, $x_a = Ae^{i\omega t}$ and $x_b = Be^{i\omega t}$. Substituting for x_a and x_b into the equations of motion and dividing through by $e^{i\omega t}$ leads to

$$A(2k - m\omega^2) = Bk \qquad (4.26a)$$

and

$$Ak = B(k - m\omega^2). \qquad (4.26b)$$

Equation (4.26) leads to the quadratic equation $(m\omega^2)^2 - 3km\omega^2 + k^2 = 0$, which has the solutions $\omega^2 = (k/2m)(3 \pm \sqrt{5})$, giving the two normal frequencies. Substituting for $\omega^2 = (k/2m)(3 - \sqrt{5})$ in Equation (4.26a) gives $A/B = 1/2(\sqrt{5} - 1)$, while substituting for $\omega^2 = (k/2m)(3 + \sqrt{5})$ gives $A/B = -1/2(\sqrt{5} + 1)$ where the minus sign indicates that the masses move in opposite directions, i.e. in anti-phase.

A powerful way to handle the simultaneous equations that arise for coupled oscillators is to use a matrix representation.[1] This works as follows for the example above. Equation (4.26) can be written, respectively, as

$$\frac{2k}{m}A - \frac{k}{m}B = \omega^2 A, \qquad (4.27a)$$

[1] This matrix approach can be omitted by the reader without detriment, although it is extremely powerful in more complicated cases.

and

$$-\frac{k}{m}A + \frac{k}{m}B = \omega^2 B. \tag{4.27b}$$

In matrix form these equations become

$$\begin{bmatrix} \dfrac{2k}{m}, & -\dfrac{k}{m} \\ -\dfrac{k}{m}, & \dfrac{k}{m} \end{bmatrix} \begin{bmatrix} A \\ \\ B \end{bmatrix} = \omega^2 \begin{bmatrix} A \\ \\ B \end{bmatrix}. \tag{4.28}$$

This is an *eigenvalue* equation. The solutions of this equation for ω^2 are called the *eigenvalues*. The column vector with components A and B is an *eigenvector* of the matrix. We can rewrite Equation (4.28) in the following form

$$\begin{bmatrix} \left(\dfrac{2k}{m} - \omega^2\right), & -\dfrac{k}{m} \\ -\dfrac{k}{m}, & \left(\dfrac{k}{m} - \omega^2\right) \end{bmatrix} \begin{bmatrix} A \\ \\ B \end{bmatrix} = 0. \tag{4.29}$$

This equation has non-zero solutions if and only if the determinant vanishes, i.e. if

$$\left(\frac{2k}{m} - \omega^2\right)\left(\frac{k}{m} - \omega^2\right) - \left(\frac{k}{m}\right)^2 = 0,$$

giving $m^2\omega^4 - 3km\omega^2 + k^2 = 0$ and the solutions $\omega^2 = (k/2m)(3 \pm \sqrt{5})$ as before. Substituting for these solutions in Equation (4.28) yields the two values of A/B. The power of this approach is not obvious for the case of two coupled oscillators but it quickly becomes apparent when more than two are involved.

In this section we have discussed the example of two masses connected by springs where the masses oscillate in one dimension, i.e. along the x-axis. We found that this system has two normal modes of oscillation and that each mode has an associated normal coordinate q and normal angular frequency ω. These results can be generalised to N masses interconnected by springs and moving in three dimensions. As for the case of two masses the N masses do not move independently. When one mass is set oscillating the other masses will feel the disturbance and will start to oscillate. For N coupled masses there are $3N$ normal modes of oscillation where the factor of 3 corresponds to the three perpendicular directions along which each mass can move. Again each normal mode has a normal coordinate and normal frequency, so that we have normal coordinates $q_1, q_2, \ldots, q_{3N}$ with corresponding normal frequencies $\omega_1, \omega_2, \ldots, \omega_{3N}$. For each normal mode we have independent SHM in the coordinate q with frequency ω. A good example of this is provided by a crystal lattice. In Section 1.2.6 we described how an atom in a crystal can be modelled as a simple harmonic oscillator and how Einstein used this model to explain the variation of the specific heat of a crystal with temperature. Although Einstein's model had great success in explaining the main features of this behaviour, the model is a great oversimplification and has limitations. This is because it assumes that the atoms vibrate totally independently of each other about fixed lattice sites. In reality, they do not because the atoms are coupled together. A macroscopic mechanical analogue of a crystal lattice would consist of billiard balls connected together with identical springs. Figure 4.13 shows a two-dimensional picture of this. If one ball is set vibrating, say the one labelled A in Figure 4.13, a disturbance will propagate throughout the whole system until all the balls are vibrating. Similarly, the atoms in a crystal are coupled rather than independent oscillators. Einstein's theory can be improved by describing the N atoms in a crystal in terms of the $3N$ normal modes of vibration of the whole crystal, each with its own characteristic angular frequency $\omega_1, \omega_2, \ldots, \omega_{3N}$. In terms of these normal modes, the lattice vibrations are equivalent to $3N$ independent harmonic oscillators with these angular frequencies (see also Mandl,[2] Section 6.3).

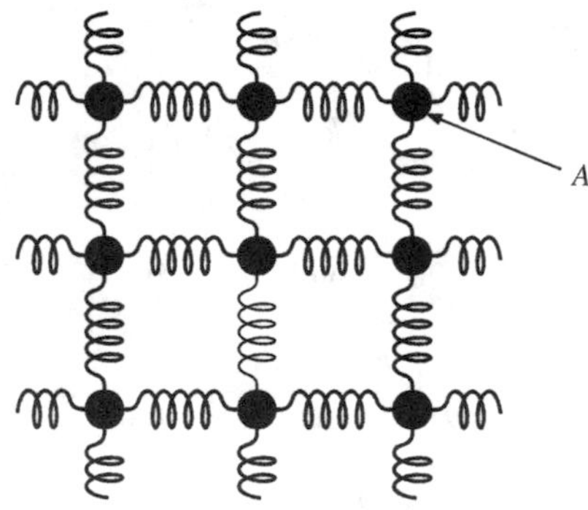

Figure 4.13 Two-dimensional analogue of a crystal lattice, consisting of billiard balls connected by springs.

Coupling can also occur in oscillating electrical circuits (cf. Figure 1.21). An electrical version of a coupled oscillator is shown in Figure 4.14. A mutual (shared) inductor M couples together the two electrical circuits where the magnetic flux arising from the current in one circuit threads the second circuit. Any change of flux induces a voltage in both circuits. A transformer, which is used to change the amplitude of an AC voltage, depends upon mutual inductance for its operation.

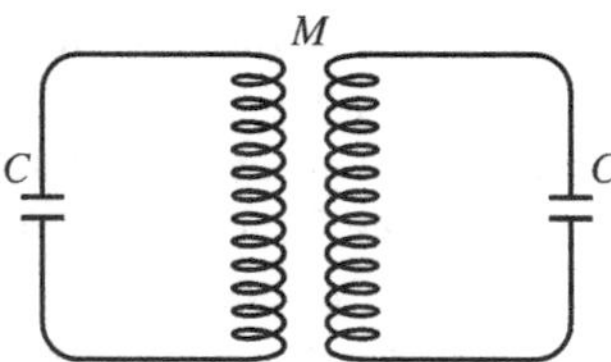

Figure 4.14 Example of a coupled electrical oscillator, where the coupling is provided by the mutual inductance M.

[2] Statistical Physics, F. Mandl, Second Edition, 1988, John Wiley & Sons, Ltd.

4.5 FORCED OSCILLATIONS OF COUPLED OSCILLATORS

We saw in Chapter 3 that the amplitude of oscillation of a harmonic oscillator becomes very large when a periodic driving force is applied at its natural frequency of oscillation. At other driving frequencies the amplitude is relatively small. For the case of two oscillators coupled together we may expect similar behaviour. Now, however, there are two natural frequencies corresponding to the two normal frequencies. Thus we may expect that the system will exhibit large amplitude oscillations when the driving frequency is close to either of these two normal frequencies. This is indeed the case. We can explore forced oscillations by considering the arrangement of two masses connected by springs as shown in Figure 4.15. This is similar to the arrangement shown in Figure 4.10 but now the end s of one of the outer springs is moved harmonically as $\xi = a\cos\omega t$. The displacements ξ, x_a and x_b of the masses from equilibrium are shown in Figure 4.15 at some instant

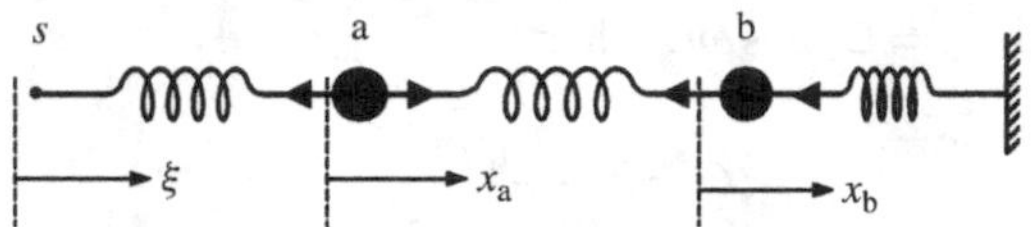

Figure 4.15 Forced oscillations of a coupled oscillator. The end s of the spring is moved harmonically as $\xi = a\cos\omega t$.

of time. The resulting equation of motion for mass a is

$$m\frac{d^2x_a}{dt^2} = -k(x_a - \xi) + k(x_b - x_a) \tag{4.30}$$

giving

$$\frac{d^2x_a}{dt^2} + \frac{2k}{m}x_a - \frac{k}{m}x_b = \frac{F_0}{m}\cos\omega t, \tag{4.31}$$

where $F_0 = ka$. Similarly, the equation of motion for mass b is

$$\frac{d^2x_b}{dt^2} - \frac{k}{m}x_a + \frac{2k}{m}x_b = 0. \tag{4.32}$$

We can solve these two simultaneous equations by, respectively, adding and subtracting them. Thus

$$\frac{d^2(x_a + x_b)}{dt^2} + \frac{k}{m}(x_a + x_b) = \frac{F_0}{m}\cos\omega t \tag{4.33}$$

and

$$\frac{d^2(x_a - x_b)}{dt^2} + \frac{3k}{m}(x_a - x_b) = \frac{F_0}{m}\cos\omega t. \tag{4.34}$$

We now change variables to the normal coordinates

$$q_1 = (x_a + x_b) \text{ and } q_2 = (x_a - x_b) \tag{4.35}$$

giving

$$\frac{d^2q_1}{dt^2} + \frac{k}{m}q_1 = \frac{F_0}{m}\cos\omega t \tag{4.36}$$

and

$$\frac{d^2q_2}{dt^2} + \frac{3k}{m}q_2 = \frac{F_0}{m}\cos\omega t. \tag{4.37}$$

This is a striking result and illustrates the power and simplicity of describing the coupled motion in terms of the normal coordinates. For each of the *independent* coordinates q_1 and q_2 we have the equation for forced oscillations of a simple harmonic oscillator, i.e. an equation of the same form as Equation (3.1) in Section 3.2.1, and we can at once take over the solutions, Equations (3.5a) and (3.7a), from that section. We can describe the steady state solutions by the equations $q_1 = C_1\cos\omega t$ and $q_2 = C_2\cos\omega t$, where

$$C_1 = \frac{F_0/m}{(\omega_1^2 - \omega^2)}, \tag{4.38}$$

$$C_2 = \frac{F_0/m}{(\omega_2^2 - \omega^2)} \tag{4.39}$$

and where $\omega_1^2 = k/m$ and $\omega_2^2 = 3k/m$. The maximum values of C_1 and C_2 given by these equations are infinitely large when $\omega = \omega_1$ and $\omega = \omega_2$, respectively, so that the amplitudes of oscillation would become infinite if the system were driven at one of its normal frequencies. (We had a similar situation when considering a driven oscillator in Section 3.2.1.) This is, of course, because we have neglected damping that would limit their values in real situations. Nevertheless we can conclude that a coupled oscillator will oscillate with large amplitude when it is driven at either of its normal frequencies. At other driving frequencies the masses will oscillate at the driving frequency but with much smaller amplitude. From Equation (4.35) we have

$$x_a = \frac{1}{2}(q_1 + q_2) = \frac{1}{2}(C_1 + C_2)\cos\omega t$$

and

$$x_b = \frac{1}{2}(q_1 - q_2) = \frac{1}{2}(C_1 - C_2)\cos\omega t.$$

It follows from Equations (4.38) and (4.39) that when the driving frequency ω is near the first normal frequency $\omega_1 = \sqrt{k/m}$, we have $|C_1| \gg |C_2|$, and $x_a \approx x_b$, i.e. the two masses oscillate in phase. When the driving frequency ω is near the second normal frequency $\omega_2 = \sqrt{3k/m}$, one similarly obtains $x_a \approx -x_b$, i.e. the two masses oscillate in anti-phase.

Since a coupled system oscillates with large amplitude when driven at one of its normal frequencies this provides a way of determining these frequencies experimentally. A good example of this is provided by the vibrations of molecules that contain more than two atoms. For example, the molecule carbon dioxide (CO_2) can be modelled by three masses connected by two springs in a linear configuration (see Figure 4.16). The central mass represents the carbon atom and the other two masses represent the oxygen atoms while the springs represent the molecular bonds. This system has two normal modes of vibration for displacements along the line connecting the masses. These are called the *symmetric stretch* mode and the *asymmetric stretch* mode as illustrated in Figure 4.16(a) and (b), respectively. In the symmetric stretch mode the central mass remains fixed in position while the two outer masses vibrate against it. In the asymmetric stretch mode the two outer masses move in the same direction and maintain the same distance apart. However, since there is no net translational motion, the central mass moves in the opposite direction to keep the position of the centre of mass stationary. The normal frequencies of molecular vibrations are determined experimentally by *absorption spectroscopy*. In this technique, radiation of tunable frequency is passed through a cell containing the molecules of interest. The oscillating electric field of the radiation interacts with the molecule, which behaves like a driven oscillator (see also Section 3.3). The intensity of the radiation, after it has passed through the cell, is measured as a function of its frequency. This gives the *absorption spectrum* of the molecule. When the frequency of the radiation matches a normal frequency, the radiation is strongly absorbed by the molecules. (We are effectively observing the power resonance curve, see also Section 3.3.) The frequencies at which this absorption occurs give directly the normal mode frequencies of the molecule. The measured values of the frequency ν for the symmetric stretch and the asymmetric stretch modes of the CO_2 molecule are $4.0 \times 10^{13}\ s^{-1}$ and $7.0 \times 10^{13}\ s^{-1}$, respectively. The CO_2 molecule also has a *bending* mode of vibration as illustrated in Figure 4.16(c). The frequency of this mode is $2.0 \times 10^{13}\ s^{-1}$. This bending motion can occur in two orthogonal planes and since these have the same frequency of vibration they are said to be *degenerate* in frequency. These frequencies lie in the far infrared region of the electromagnetic spectrum, with corresponding wavelengths of $\sim 10\ \mu m$.

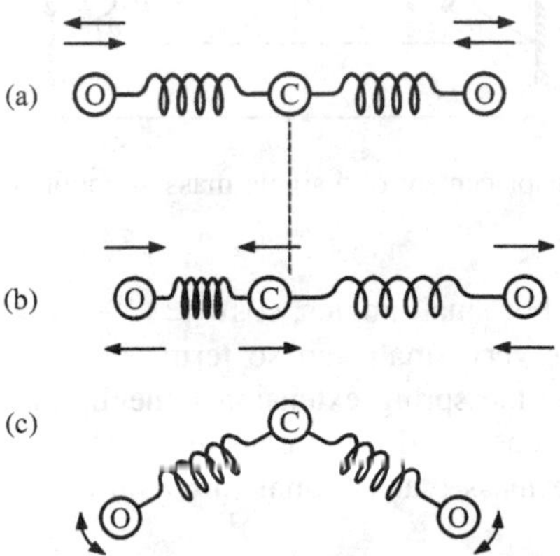

Figure 4.16 A model of the normal modes of vibration of the CO_2 molecule: (a) the symmetric stretch mode; (b) the asymmetric stretch mode; and (c) the bending mode.

Vibrations of CO_2 molecules and some other molecules in the Earth's atmosphere play a key role in *global warming* because they strongly absorb radiation in the far infrared. The surface temperature of the Sun is 5800 K and the radiation emitted by the Sun peaks at about 500 nm. However, the surface of the Earth is at a much lower temperature, ~300 K, and its radiation peaks at ~10 μm. The Earth's atmosphere is largely transparent at visible and near infrared wavelengths and the Sun's radiation passes through. However, the global-warming molecules absorb the Earth's far infrared radiation and act to trap its energy. This effect leads to an increase in the surface temperature of the Earth.

4.6 TRANSVERSE OSCILLATIONS

In our discussion of the oscillations of masses coupled by springs (Section 4.4) the periodic displacements of the masses took place along a line connecting them. These are called *longitudinal* oscillations. It is also possible to have periodic displacements in a direction perpendicular to this line. These are called *transverse* oscillations and will be discussed further in Chapter 5. In the meantime we will first consider the transverse oscillations of a single mass m connected by two springs as shown in Figure 4.17. These have a spring constant k, and the length l of each spring is greater than the unstretched length so that there is a tension T in the springs. The mass is displaced in the transverse direction by a distance y, where upward displacements are taken as positive. We first note that for small displacements the tension in the springs remains constant, which we can see as follows. For a displacement y, each spring will be extended by an amount Δl given by

$$\Delta l = l\left(\frac{1}{\cos\theta} - 1\right)$$

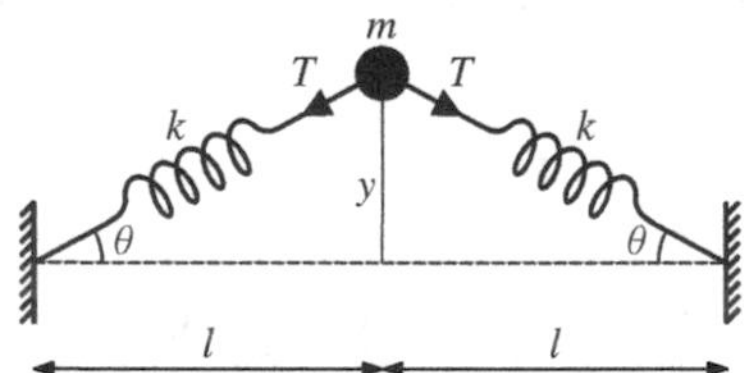

Figure 4.17 Transverse displacement of a single mass m coupled by two springs of spring constant k.

where $\theta = \arctan(y/l)$. For small angles, $\cos\theta \simeq (1-\theta^2)^{1/2}$, and so $\Delta l \simeq l\theta^2/2$. If θ is small then θ^2 is very small and so terms in θ^2 can be neglected. Then to a good approximation the spring extension is negligibly small and the tension in the spring T can be considered to be constant. The springs do, however, exert a restoring force on the mass that is equal to $2T\sin\theta$. The resultant equation of motion is

$$m\frac{d^2y}{dt^2} = -2T\sin\theta \simeq -2T\theta \simeq -2T\frac{y}{l} \tag{4.40}$$

for small θ, giving to a good approximation:

$$\frac{d^2y}{dt^2} = -\frac{2T}{ml}y \tag{4.41}$$

This is the equation of SHM with frequency $\sqrt{2T/ml}$. The system has this one normal mode of vibration.

We now extend our discussion to a coupled oscillator consisting of two equal masses connected by three identical springs of length l and under tension T, as shown in Figure 4.18. The masses are displaced in the transverse direction by distances of y_a and y_b, respectively. The directions of the forces acting on the masses are indicated by the arrow heads and the resultant equations of motion for the two masses are derived as follows. For mass a, we have

$$m\frac{d^2y_a}{dt^2} = -T\sin\theta_1 + T\sin\theta_2 \tag{4.42}$$

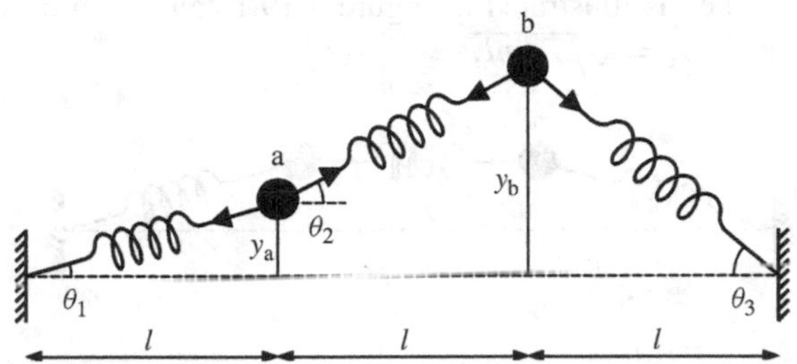

Figure 4.18 Transverse displacements of two masses connected by springs.

giving, for small displacements,

$$m\frac{d^2y_a}{dt^2} = -\frac{T}{l}y_a + \frac{T}{l}(y_b - y_a) = \frac{T}{l}(y_b - 2y_a). \tag{4.43}$$

Similarly, we have for mass b

$$m\frac{d^2y_b}{dt^2} = -T\sin\theta_2 - T\sin\theta_3$$

giving

$$m\frac{d^2y_b}{dt^2} = \frac{T}{l}(y_a - 2y_b). \tag{4.44}$$

Substituting $y_a = Ae^{i\omega t}$ and $y_b = Be^{i\omega t}$ into Equations (4.43) and (4.44) and dividing through by $e^{i\omega t}$ leads to

$$A\left(\frac{2T}{l} - m\omega^2\right) = \frac{T}{l}B \tag{4.45}$$

and

$$\frac{T}{l}A = B\left(\frac{2T}{l} - m\omega^2\right). \tag{4.46}$$

Equations (4.45) and (4.46) give two expressions for A/B, and equating these leads to the quadratic equation in ω^2:

$$\left(\frac{2T}{l} - m\omega^2\right)^2 = \left(\frac{T}{l}\right)^2 \tag{4.47}$$

with the solutions $\omega^2 = T/ml$ and $3T/ml$. Substituting for $\omega^2 = T/ml$ in Equation (4.45) gives $A = B$. This corresponds to the first normal mode of the system where both masses move in the same directions as each other as illustrated in Figure 4.19(a) and each performs SHM at the normal frequency $\omega_1 = \sqrt{T/ml}$. Substituting for $\omega^2 = 3T/ml$ in Equation (4.45) gives $A = -B$. This corresponds to the second normal mode of the system where the two masses move in opposite directions to each other as illustrated in Figure 4.19(b) and each performs SHM at the normal frequency $\omega_2 = \sqrt{3T/ml}$.

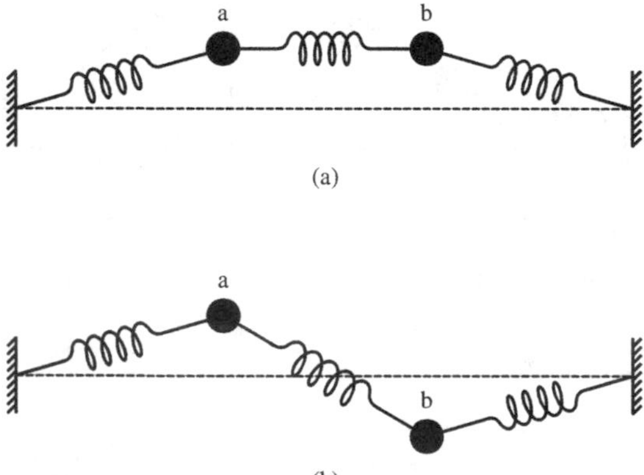

Figure 4.19 The two normal modes for transverse oscillations of two masses connected by springs where (a) the masses move in the same directions as each other and (b) they move in opposite directions.

We see that the frequency of oscillation depends on the particular normal mode. It is also proportional to the square root of the tension T and inversely proportional to the square root of the mass m. We will encounter similar relationships for standing waves on taut strings in Chapter 5. Indeed the normal modes shown in Figure 4.19 are already starting to resemble standing waves on a taut string. This similarity is even more striking when we have a larger number N of masses. To emphasise this similarity we show in Figure 4.20 an arrangement of nine masses connected by elastic strings of equal length l. The figure shows schematically three of the possible modes of oscillation of this arrangement. Without pursuing the details of the mathematics, we note that in each normal mode all the individual masses oscillate in SHM at the same frequency, equal to the normal frequency. The amplitude of the oscillations will, however, vary from mass to mass as indicated by Figure 4.20. The number of normal modes is equal to the number of masses and the highest possible normal mode will occur when alternate masses are moving in opposite directions, as shown in Figure 4.20(c). This gives an upper limit to the highest normal frequency that is possible. If we simultaneously take the limits $N \to \infty, m \to 0$ and $l \to 0$, in such a way that Nm remains finite, we indeed obtain the situation of standing waves on a taut string. Thus we see that coupled oscillators are the bridge between vibrations and waves. Our discussion of coupled oscillators has also seen the repeated appearance of SHM again and again, and this further emphasises the importance and diversity of this form of motion.

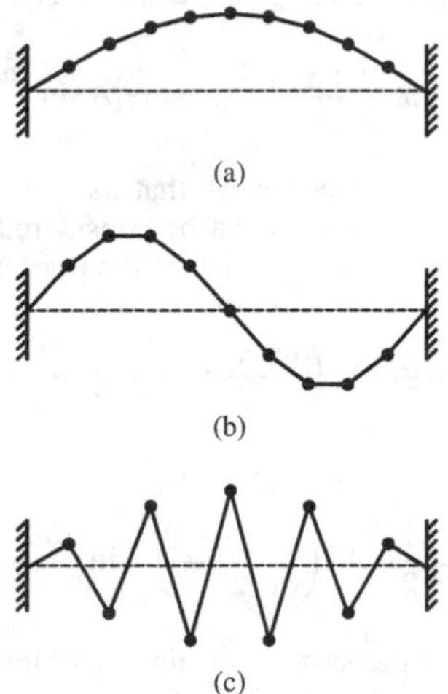

Figure 4.20 Some normal modes of transverse oscillations for nine masses connected by elastic strings: (a) the first normal mode; (b) the second normal mode; and (c) the highest normal mode.

PROBLEMS 4

4.1 Two simple pendulums, each of length 0.300 m and mass 0.950 kg, are coupled by attaching a light, horizontal spring of spring constant $k = 1.50$ N m^{-1} to the masses. (a) Determine the frequencies of the two normal modes. (b) One of the pendulums is held at a small distance away from its equilibrium position while the other pendulum is held at its equilibrium position. The two pendulums are then released simultaneously. Show that after a time of approximately 12 s the amplitude of oscillation of the first pendulum will become equal to zero momentarily. (Assume $g = 9.81$ m s^{-2}.)

4.2 Two simple pendulums, each of length 0.50 m and mass 5.0 kg, are coupled by attaching a light, horizontal spring of spring constant $k = 20$ N m^{-1} to the masses. (a) One of the masses is held at a horizontal displacement $x_a = +5.0$ mm while the other mass is held at a horizontal displacement $x_b = +5.0$ mm. The two masses are then released from rest simultaneously. Using the expressions

$$x_a = \frac{1}{2}(C_1 \cos \omega_1 t + C_2 \cos \omega_2 t) \text{ and } x_b = \frac{1}{2}(C_1 \cos \omega_1 t - C_2 \cos \omega_2 t)$$

where ω_1 and ω_2 are the normal frequencies, find the values of C_1 and C_2. Plot x_a and x_b as a function of time t over the time interval $t = 0$ to 10 s. (b) Repeat part (a) for initial conditions: (i) $x_a = +5.0$ mm, $x_b = -5.0$ mm, (ii) $x_a = +10$ mm, $x_b = 0$ mm and (iii) $x_a = +10$ mm, $x_b = +5.0$ mm. (Assume $g = 9.81$ m s^{-2}.)

4.3 Consider the example of two identical masses connected by three identical springs as shown in Figure 4.11. Combine the equations of motion of the two masses to obtain a pair of equations of the form

$$\frac{d^2 q_1}{dt^2} + \omega_1^2 q_1 = 0 \text{ and } \frac{d^2 q_2}{dt^2} + \omega_2^2 q_2 = 0$$

and hence obtain the normal coordinates q_1 and q_2 and the respective normal frequencies ω_1 and ω_2.

4.4 Two identical pendulums of the same mass m are connected by a light spring. The displacements of the two masses are given, respectively, by

$$x_a = A\cos\frac{(\omega_2 - \omega_1)t}{2}\cos\frac{(\omega_2 + \omega_1)t}{2}, \quad x_b = A\sin\frac{(\omega_2 - \omega_1)t}{2}\sin\frac{(\omega_2 + \omega_1)t}{2}.$$

Assume that the spring is sufficiently weak that its potential energy can be neglected and that the energy of each pendulum can be considered to be constant over a cycle of its oscillation. (a) Show that the energies of the two masses are

$$E_a = \frac{1}{2}mA^2\left(\frac{\omega_2 + \omega_1}{2}\right)^2\cos^2\frac{(\omega_2 - \omega_1)t}{2}$$

and

$$E_b = \frac{1}{2}mA^2\left(\frac{\omega_2 + \omega_1}{2}\right)^2\sin^2\frac{(\omega_2 - \omega_1)t}{2}$$

and that the total energy of the system remains constant. (b) Sketch E_a and E_b over several cycles on the same graph. What is the frequency at which there is total exchange of energy between the two masses?

4.5

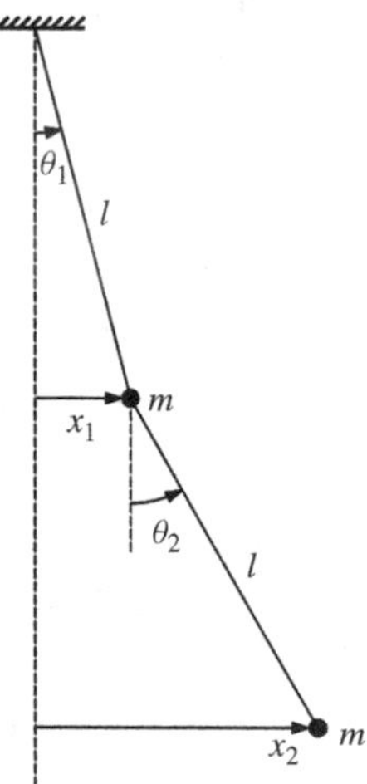

Two identical masses of mass m are suspended from a rigid support by two strings of length l and oscillate in the vertical plane as illustrated by the figure. The oscillations are of sufficiently small amplitude that any changes in the tensions of the two strings from their values when the system is in static equilibrium can be neglected. In addition the small-angle approximation $\sin\theta \simeq \theta$ can be made. (a) Show that the equations of motions of the upper and lower masses, respectively, are

$$\frac{\mathrm{d}^2 x_1}{\mathrm{d}t^2} + \frac{3g}{l}x_1 - \frac{g}{l}x_2 = 0$$

and

$$\frac{\mathrm{d}^2 x_2}{\mathrm{d}t^2} + \frac{g}{l}x_2 - \frac{g}{l}x_1 = 0.$$

(b) Assuming solutions of the form $x_1 = A\cos\omega t$ and $x_2 = B\cos\omega t$, show that the two normal frequencies of the system are $\sqrt{(2 \pm \sqrt{2})g/l}$ and find the corresponding ratios, B/A. (c) Determine the periods of the two normal modes for $l = 1.0$ m and compare these with the period of a simple pendulum of this length. (Assume $g = 9.81$ m s^{-2}.)

4.6

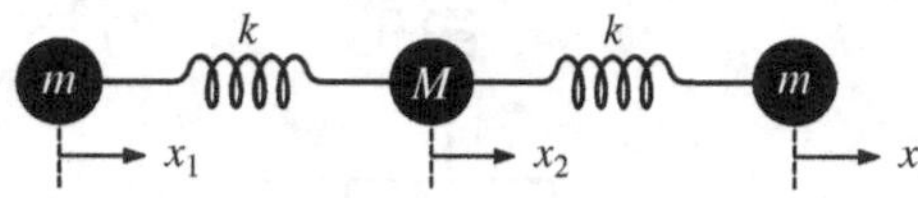

The figure shows two identical masses of mass m connected to a third mass of mass M by two identical springs of spring constant k. Consider vibrations of the masses along the line joining their centres where x_1, x_2 and x_3 are their respective displacements from equilibrium. (a) Without any mathematical detail, use your physical intuition to deduce the normal frequency for *symmetric-stretch* vibrations. (b) Show that the equations of motion of the three masses are:

$$\frac{\mathrm{d}^2 x_1}{\mathrm{d}t^2} + \omega_1^2 x_1 - \omega_1^2 x_2 = 0,$$

$$\frac{\mathrm{d}^2 x_2}{\mathrm{d}t^2} - \omega_2^2 x_1 + 2\omega_2^2 x_2 - \omega_2^2 x_3 = 0$$

and

$$\frac{\mathrm{d}^2 x_3}{\mathrm{d}t^2} - \omega_1^2 x_2 + \omega_1^2 x_3 = 0,$$

where $\omega_1^2 = k/m$ and $\omega_2^2 = k/M$. (c) Show that the normal frequencies of the system are $\sqrt{k/m}$ and $\sqrt{k(2m+M)/Mn}$. (d) Determine the ratio of normal frequencies for $m/M = 16/12$ and compare with the vibrational frequencies of the CO_2 molecule given in the text.

4.7

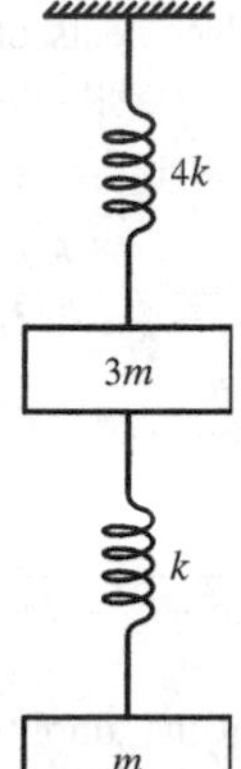

The figure shows two masses of mass $3m$ and m hanging from springs of spring constants $4k$ and k, respectively. (a) Show that the normal frequencies of oscillation are $\sqrt{2k/m}$ and $\sqrt{2k/3m}$. (b) Describe the two normal modes.

4.8

Five identical masses are connected by six identical springs between two rigid walls, as illustrated in the figure, and move without friction on a horizontal surface. How many normal modes of vibration in the *transverse* direction does the system have? Sketch these normal modes bearing in mind that the transverse positions of the masses pass through sinusoidal curves (cf. Figure 4.20).

4.9

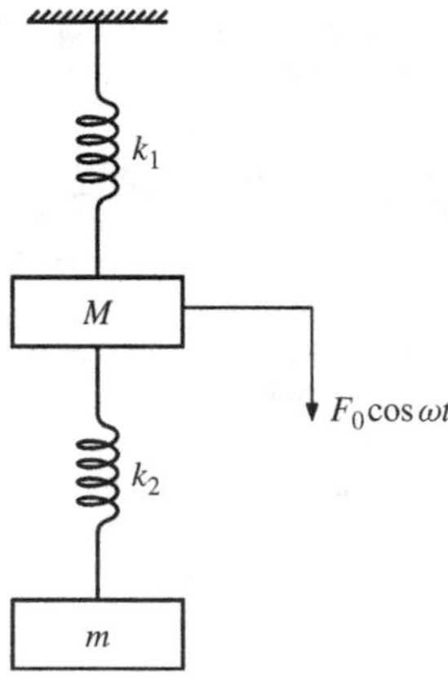

The figure shows two masses M and m suspended from a rigid ceiling by springs of spring constant k_1 and k_2. (a) If the mass M is subjected to a driving force $F_0 \cos \omega t$ in the downward direction, show that the equations of motion of the masses are

$$M\frac{\mathrm{d}^2 x_1}{\mathrm{d}t^2} + (k_1 + k_2)x_1 - k_2 x_2 = F_0 \cos \omega t$$

and

$$m\frac{\mathrm{d}^2 x_2}{\mathrm{d}t^2} - k_2 x_1 + k_2 x_2 = 0,$$

where x_1 and x_2 are the displacements of the masses M and m, respectively, from their equilibrium positions. (b) Assuming solutions of the form $x_1 = A\cos \omega t$ and $x_2 = B\cos \omega t$ show that

$$A = \frac{F_0(k_2 - m\omega^2)}{(k_1 + k_2 - M\omega^2)(k_2 - m\omega^2) - k_2^2}$$

and

$$B = \frac{F_0 k_2}{(k_1 + k_2 - M\omega^2)(k_2 - m\omega^2) - k_2^2}.$$

(c) For $\omega = \sqrt{k_1/M}$ show that the amplitude of vibration of mass M will be zero if $k_2/k_1 = m/M$.

4.10

Three identical masses of mass m are connected by four identical springs of spring constant k between two rigid walls, as shown in the figure, and move without friction on a horizontal surface. They vibrate along the line joining their centres. (a) Show that the normal frequencies of the system are $\sqrt{2k/m}$ and $\sqrt{(2 \pm \sqrt{2})k/m}$. (b) Describe the three normal modes of vibration.

[Hint: The determinant $\begin{vmatrix} a_{11} & a_{12} & a_{13} \\ a_{21} & a_{22} & a_{23} \\ a_{31} & a_{32} & a_{33} \end{vmatrix} =$

$a_{11}(a_{22}a_{33} - a_{32}a_{23}) + a_{21}(a_{32}a_{13} - a_{12}a_{33}) + a_{31}(a_{12}a_{23} - a_{22}a_{13})$.]

5

Travelling Waves

Waves arise in a wide range of physical phenomena. They occur as ripples on a pond and as seismic waves following an earthquake. Music is carried by sound waves and most of what we know about the Universe comes from electromagnetic waves that reach the Earth. Furthermore, we communicate with each other through a variety of different waves. At the microscopic level, the particles of matter have a wave nature as expressed by quantum wave mechanics. At the other end of the scale, scientists are trying to detect gravitational waves that are predicted to occur when massive astronomical objects like black holes move rapidly. Even a Mexican wave travelling around a sports arena has many of the characteristics of wave motion. It is not surprising therefore that waves are at the heart of many branches of the physical sciences including optics, electromagnetism, quantum mechanics and acoustics.

In this chapter we begin to explore the physical characteristics of waves and their mathematical description. We distinguish between *travelling waves* and *standing waves*. Ripples on a pond are an example of travelling waves. A plucked guitar string provides an example of a standing wave. The present chapter is devoted to travelling waves while standing waves will be discussed in Chapter 6. Travelling waves may be either *transverse waves* or *longitudinal waves*. We have already seen the difference between these two types of motion in Chapter 4. In transverse waves the change in the corresponding physical quantity, e.g. displacement, occurs in the direction at right angles to the direction of travel of the wave, as for the outgoing ripples on a pond. For longitudinal waves, the change occurs along the direction of travel. An example of this is the longitudinal compressions and rarefactions of the air that occur in the propagation of a sound wave. It is easier to see the physical processes going on in a transverse wave and so we will concentrate on them in the present chapter. However, both transverse and longitudinal waves are solutions of the *wave equation*, which is one of the most fundamental equations in physics. We will deal with *mechanical waves* that travel through some material or *medium*. However, not all waves are mechanical waves: electromagnetic waves can propagate even in a vacuum. We will discuss the energy carried by a wave

Vibrations and Waves George C. King

and the behaviour of a wave when it encounters a boundary in passing from one medium to another. Most of our discussion will be devoted to waves travelling in one dimension but we will introduce waves that move in two or three dimensions. These have much in common with one-dimensional waves.

5.1 PHYSICAL CHARACTERISTICS OF WAVES

When we observe a wave it is clear that something, that we may call a disturbance, travels or propagates from one region of a medium to another. This disturbance travels at a definite velocity v that is usually determined by the mechanical properties of the medium. For a taut string these are the mass per unit length and the tension in the string. However, the medium does not travel with the wave. For example, if we tap one end of a solid metal rod, a sound wave propagates along the rod but the rod itself does not travel with the wave. (For this reason, waves can travel at high velocities.) In fact, the particles of the rod move about their equilibrium positions to which they are bound. We saw such behaviour for the transverse oscillations of masses connected by springs in Section 4.6. There the equilibrium position was the straight line along which the masses lie when at rest and the springs provide the restoring force. We also saw in Chapter 4 that an oscillator can transfer all of its energy to another oscillator to which it is coupled under appropriate conditions. A simplified picture of a wave travelling through a medium is therefore a long line of oscillators coupled together in some way, just like the atoms in a one-dimensional crystal. Then if the end oscillator is displaced from its equilibrium position it exerts a force on its neighbour. In turn this force and the resultant displacement propagate down the line of oscillators. Energy must be put into the system to cause the initial disturbance and it is this energy that is transmitted by the wave. This energy is evident as the destructive power of a tsunami and in the warmth of the Sun's rays. On a sunny day the solar energy deposited on the Earth's surface is about 1 kJ m^{-2} s^{-1}; a power of 1 kW m^{-2}. This is a substantial amount of power that is an increasingly important source of energy for the World's needs.

5.2 TRAVELLING WAVES

A common experience is to take the end of a long rope like a clothesline and move one end of it up and down rapidly to launch a wave pulse down the rope. A schematic diagram of this is shown in Figure 5.1. The pulse roughly holds its shape and travels with a definite velocity along the rope. Here we will use a *Gaussian function* to model this travelling wave pulse. The Gaussian function can be represented by

$$y = A\exp[-(x^2/a^2)], \tag{5.1}$$

where A and a are constants. This function appears in many branches of the physical sciences and is plotted in Figure 5.2. When $x = 0$, $y = A$ and when $x = \pm a$,

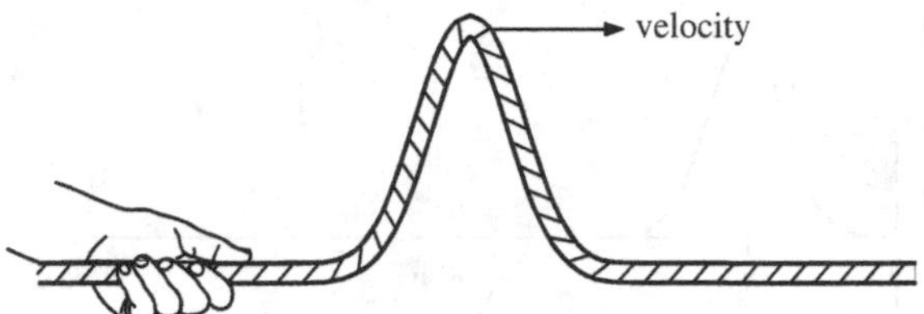

Figure 5.1 A wave pulse can be launched down a long rope by moving the end of the rope rapidly up and down.

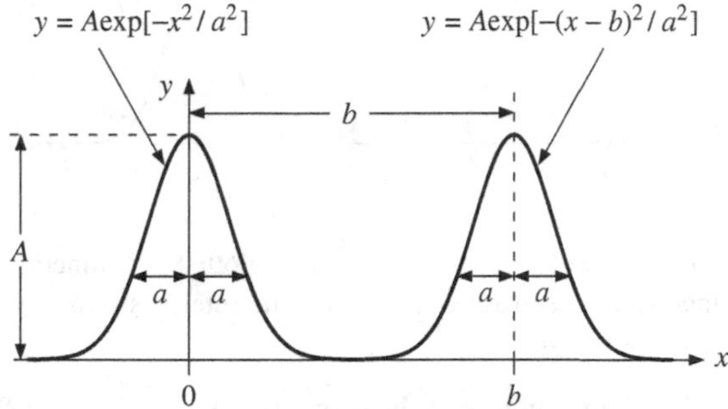

Figure 5.2 The Gaussian functions $y = A\exp[-(x^2/a^2)]$ and $y(x) = A\exp[-(x-b)^2/a^2]$. A is the height of the Gaussian and a characterises its width. These two Gaussians have the same shape but are separated by distance b.

$y = A/\mathrm{e}$. A corresponds to the *height* of the Gaussian and a is a measure of its *width*. If we now change the variable x to $(x - b)$ we obtain

$$y = A\exp[-(x-b)^2/a^2]. \tag{5.2}$$

This function is also plotted in Figure 5.2. We see that the shape of the function, as characterised by its height and width, is the same as before. We have simply moved the Gaussian a distance b to the left, so that now it has its maximum value A at $x = b$. Suppose we now change the variable x to $(x - vt)$ where t is time and v is a constant with the dimensions of distance/time. Then we obtain

$$y(x,t) = A\exp[-(x-vt)^2/a^2]. \tag{5.3}$$

The value of vt increases linearly with time. Consequently, Equation (5.3) describes a Gaussian that moves in the positive x-direction at a constant rate just like the wave pulse on the rope. This is illustrated in Figure 5.3 where the Gaussian is plotted at three different instants of time that are separated by equal time intervals of δt. The rate at which it moves is the velocity v.

We can generalise the above by saying that when a wave is going in the positive x-direction, the dependence of the shape of the rope on x and t must be of

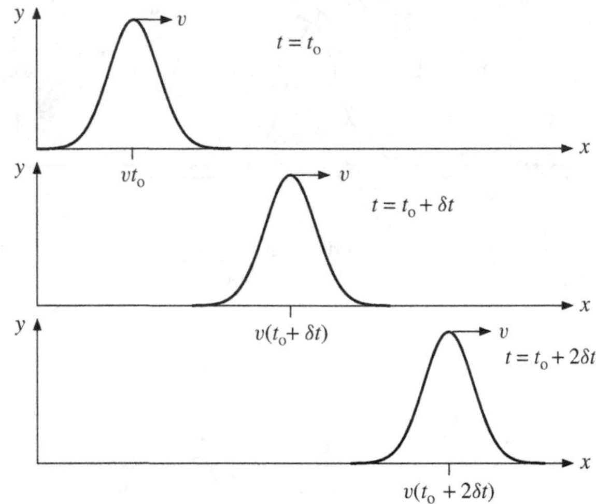

Figure 5.3 The Gaussian $y = A\exp[-(x - vt)^2/a^2]$ plotted as a function of position x, at three different instants of time, separated by equal time intervals of δt.

the general form $f(x - vt)$, where f is some function of $(x - vt)$. Examples of $f(x - vt)$ are the Gaussian function $A\exp[-(x - vt)^2/a^2]$ that we saw above, and the travelling sinusoidal wave $A\sin[2\pi(x - vt)/\lambda]$ that we will discuss in the next section. The shape of the wave is given by $f(x - vt)$ at $t = 0$, i.e. by $f(x)$ as illustrated in Figure 5.4(a). At time t, the wave has moved a distance vt to the right. However it has retained its shape, as shown in Figure 5.4(b). This is the important characteristic of wave motion: the wave retains its shape as it travels along. Clearly, we could determine the shape of the wave by taking a snapshot of the rope at a particular instant of time. However, we could also find this shape by measuring the variation in the displacement of a given point on the rope as the wave passes by. A wave travelling in the negative x-direction must be of the general form $g(x + vt)$ where g is some function of $(x + vt)$. Again at $t = 0$, $g(x)$

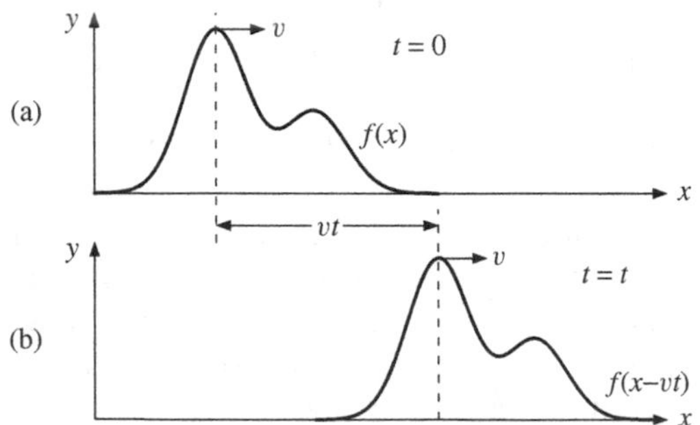

Figure 5.4 A wave travelling in the positive x-direction, defined by the function $y = f(x - vt)$. (a) $f(x - vt) \equiv f(x)$ at time $t = 0$, which gives the shape of the wave. (b) $f(x - vt)$ at time t when the wave has moved a distance vt to the right.

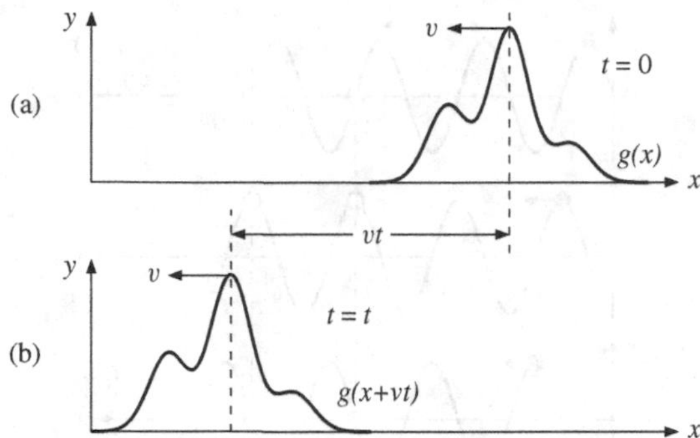

Figure 5.5 A wave travelling in the negative x-direction, defined by the function $y = g(x + vt)$. (a) $g(x + vt) \equiv g(x)$ at time $t = 0$. (b) $g(x + vt)$ at time t when the wave has moved a distance vt to the left.

gives the shape of the wave as illustrated in Figure 5.5(a). At time t, the wave has moved to the left by a distance vt but its shape remains the same, as shown in Figure 5.5(b). The general form of *any* wave motion of the rope can be written as

$$y = f(x - vt) + g(x + vt) \tag{5.4}$$

and can be considered as a superposition of two waves, each of speed v, travelling in opposite directions. In Chapter 6 we will see that the superposition of waves travelling in opposite directions is of great physical importance.

5.2.1 Travelling sinusoidal waves

Sinusoidal waves are important because they occur in many physical situations, such as in the propagation of electromagnetic radiation. They are also important because more complicated wave shapes can be decomposed into a combination of sinusoidal waves. Consequently, if we understand sinusoidal waves we can understand these more complicated waves. We return to this important principle in Chapter 6. A travelling sinusoidal wave is illustrated in Figure 5.6, at various instants of time. The dotted parts of the curves indicate that the wave extends a large distance in both directions to avoid any effects due to reflections of the wave at a fixed end. Such reflections will be discussed in Section 5.7. A sinusoidal wave is a repeating pattern. The length of one complete pattern is the distance between two successive maxima (crests), or between any two corresponding points. This repeat distance is the *wavelength* λ of the wave. The sinusoidal wave propagates along the x-direction and the displacement is in the y-direction, at right angles to the propagation direction. We could generate such a sinusoidal wave by moving the end of a long rope up and down in simple harmonic motion. The displacements lie in a single plane, i.e. in the x-y plane, and so we describe the waves as *linearly polarised* in that plane. We represent the travelling sinusoidal wave by

$$y(x, t) = A \sin \frac{2\pi}{\lambda}(x - vt) \tag{5.5}$$

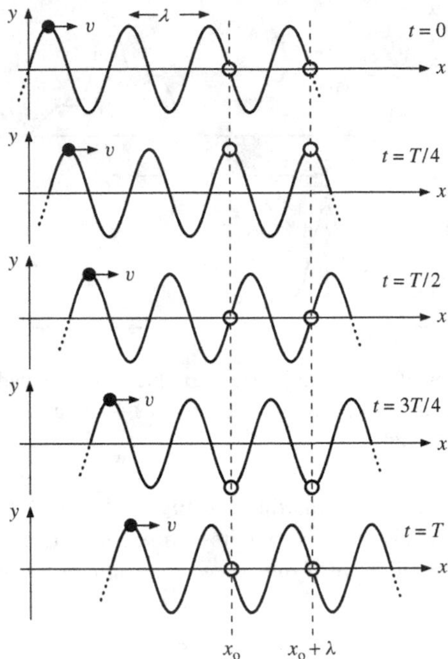

Figure 5.6 Schematic representation of a travelling sinusoidal wave of wavelength λ and period T, at the different times as indicated. Each point on the wave travels at velocity v. The open circles denote points on the wave that are separated by wavelength λ. These points move in phase with each other in the transverse direction.

where A is the amplitude and λ is the wavelength. This function repeats itself each time x increases by the distance λ. At $t = 0$, we have $y = A\sin(2\pi x/\lambda)$ which shows the sinusoidal shape of the wave. The transverse displacement y given by Equation (5.5) is a function of two variables x and t and it is interesting to see what happens if we keep either x or t fixed. Keeping x fixed is like watching a leaf on a pond that bobs up and down with the motion of the water ripples. Keeping t fixed is like taking a snapshot of the pond that fixes the positions of the water ripples in time. The sinusoidal wave travels at a definite velocity v in the positive x-direction, as can be seen from the progression of a wave crest with time in Figure 5.6. The number of times per unit time that a wave crest passes a fixed point, at say $x = x_o$, is the *frequency* ν of the wave. The frequency ν is equal to the velocity v of the wave divided by the wavelength λ. Hence we obtain

$$\boxed{\nu\lambda = v.} \tag{5.6}$$

We see that the important parameters of the wave (wavelength, frequency and velocity) are related by this simple equation. The time T that a wave crest takes to travel a distance λ is equal to λ/v, i.e. the reciprocal of the frequency. Hence,

$$\nu = \frac{1}{T}, \tag{5.7}$$

where T is the *period* of the wave.

Figure 5.6 also illustrates how the displacement of a point on the wave, at $x = x_o$, changes with time. The point moves up and down as the wave passes by and indeed its motion is simple harmonic. We can see this mathematically as follows. We have

$$y(x, t) = A \sin \frac{2\pi}{\lambda}(x - vt). \tag{5.8}$$

Then at the fixed position, $x = x_o$, we have

$$y(x_o, t) = A \sin \frac{2\pi}{\lambda}(x_o - vt). \tag{5.9}$$

Now since x has a fixed value and we want to see how y varies with t it is useful to write this equation in the equivalent form

$$y(x_o, t) = -A \sin \frac{2\pi}{\lambda}(vt - x_o), \tag{5.10}$$

using the relationship $\sin(\alpha - \beta) = -\sin(\beta - \alpha)$. Equation (5.10) shows that the displacement varies sinusoidally with time t with an *angular frequency* ω where

$$\omega = \frac{2\pi v}{\lambda} = 2\pi\nu. \tag{5.11}$$

Each point on the wave completes one period of oscillation in time period T, and we emphasise that all points along the wave oscillate at the same frequency ω. We can consider the term $2\pi x_o/\lambda$ in Equation (5.10) as a phase angle. Thus, as illustrated in Figure 5.6, points at $x = x_o$ and $x = x_o + \lambda$, denoted by the open circles, oscillate in phase with each other. As the wave propagates, any particular point on it, for example the wave crest denoted by the bold dots in Figure 5.6, maintains a constant value of transverse displacement y, and hence a constant value of $(x - vt)$. Since $(x - vt) =$ constant, $dx/dt = v$, which of course is the wave velocity.

We can use Equation (5.8) to obtain alternative mathematical expressions for the wave. Substituting for $v = \nu\lambda$ in Equation (5.8) we obtain

$$y(x, t) = A \sin\left(\frac{2\pi x}{\lambda} - 2\pi\nu t\right). \tag{5.12}$$

We define the quantity $2\pi/\lambda$ as the *wavenumber* k, i.e.

$$k = 2\pi/\lambda. \tag{5.13}$$

Substituting for $\omega = 2\pi\nu$ from Equation (5.11) and k from Equation (5.13) in Equation (5.12), we obtain

$$y(x, t) = A \sin(kx - \omega t). \tag{5.14}$$

In addition, using the relationships $\nu\lambda = v$ and $2\pi\nu = \omega$, we have

$$v = \frac{\omega}{k}. \tag{5.15}$$

The wave velocity is equal to the angular frequency divided by the wavenumber. Although we have used sine functions, we can equally well use cosine functions such as

$$y(x, t) = A\cos(kx - \omega t), \tag{5.16}$$

since the cosine function is simply the sine function with a phase difference of $\pi/2$. This is illustrated in Figure 5.7, which shows snapshots of Equations (5.14) and (5.16) at $t = 0$. We simply need to choose the solution that fits the initial conditions. Finally, in Section 3.6 we saw that it can be advantageous to use a complex representation of periodic motion. This is also the case for wave motion, remembering that, as usual, the real part of the complex form is the physical quantity. Thus we can write the following alternative mathematical expressions for travelling sinusoidal waves:

$$y(x, t) = A\exp\frac{2\pi}{\lambda}i(x - vt) \tag{5.17}$$

$$y(x, t) = A\exp 2\pi i\left(\frac{x}{\lambda} - vt\right) \tag{5.18}$$

$$y(x, t) = A\exp i(kx - \omega t). \tag{5.19}$$

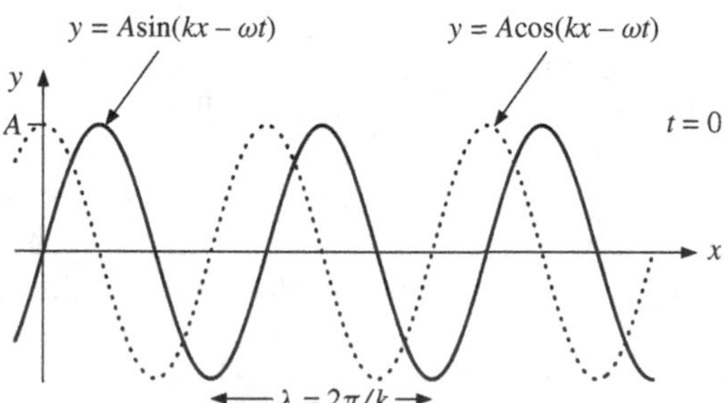

Figure 5.7 Representation of the functions $y = A\sin(kx - \omega t)$ and $y = A\cos(kx - \omega t)$ at time $t = 0$, showing the phase relationship between the two functions.

5.3 THE WAVE EQUATION

In Section 5.2 we saw that the general form of any wave motion is given by

$$y = f(x - vt) + g(x + vt). \tag{5.4}$$

We now show that this is the general solution of the *wave equation*. We start with the function $f(x - vt)$ and change variables to $u = (x - vt)$ to obtain the function $f(u)$. Notice that $f(u)$ is a function only of u. Then

$$\frac{\partial f}{\partial x} = \frac{df}{du}\frac{\partial u}{\partial x}$$

and

$$\frac{\partial^2 f}{\partial x^2} = \frac{\partial}{\partial x}\left(\frac{\mathrm{d}f}{\mathrm{d}u}\frac{\partial u}{\partial x}\right) = \frac{\mathrm{d}^2 f}{\mathrm{d}u^2}\left(\frac{\partial u}{\partial x}\right)^2 + \frac{\mathrm{d}f}{\mathrm{d}u}\left(\frac{\partial^2 u}{\partial x^2}\right).$$

Since $\partial u/\partial x = 1$ and $\partial^2 u/\partial x^2 = 0$, we have

$$\frac{\partial^2 f}{\partial x^2} = \frac{\mathrm{d}^2 f}{\mathrm{d}u^2}. \tag{5.20}$$

Similarly,

$$\frac{\partial^2 f}{\partial t^2} = v^2 \frac{\mathrm{d}^2 f}{\mathrm{d}u^2}. \tag{5.21}$$

Combining Equations (5.20) and (5.21) we obtain

$$\frac{\partial^2 f}{\partial t^2} = v^2 \frac{\partial^2 f}{\partial x^2}. \tag{5.22a}$$

Similarly, we can readily see that $g(x + vt)$ satisfies the equation

$$\frac{\partial^2 g}{\partial t^2} = v^2 \frac{\partial^2 g}{\partial x^2}. \tag{5.22b}$$

[It does not matter that the sign of the velocity has changed between $f(x - vt)$ and $g(x + vt)$ since only the square of the velocity occurs in Equation (5.22).] Thus

$$\frac{\partial^2 (f + g)}{\partial t^2} = v^2 \frac{\partial^2 (f + g)}{\partial x^2}$$

and hence we can write

$$\frac{\partial^2 y}{\partial t^2} = v^2 \frac{\partial^2 y}{\partial x^2}. \tag{5.23}$$

This is a fundamental result. Equation (5.23) is the *one-dimensional wave equation*. (The position of the velocity v in Equation (5.23) is consistent with the dimensions of the quantities involved.) The general solution of it is Equation (5.4), namely

$$y = f(x - vt) + g(x + vt). \tag{5.4}$$

The wave equation (5.23) and its general solution apply to all waves that travel in one dimension. For example, they describe sound waves in a long tube where the relevant physical parameter is the local air pressure $P(x, t)$. They describe voltage waves $V(x,t)$ on a transmission line and temperature fluctuations $T(x,t)$ along a metal rod. Consequently we write the wave equation more generally as

$$\boxed{\frac{\partial^2 \psi}{\partial t^2} = v^2 \frac{\partial^2 \psi}{\partial x^2}} \tag{5.23a}$$

and its general solution as

$$\boxed{\psi = f(x - vt) + g(x + vt),} \tag{5.4a}$$

where ψ represents the relevant physical quantity.

As a specific example of the above discussion, we have the travelling sinusoidal wave $y = A\sin[2\pi(x - vt)/\lambda]$. First, differentiating with respect to x and keeping t constant, we obtain

$$\frac{\partial y}{\partial x} = \left(\frac{2\pi}{\lambda}\right) A\cos\frac{2\pi}{\lambda}(x - vt)$$

and

$$\frac{\partial^2 y}{\partial x^2} = -\left(\frac{2\pi}{\lambda}\right)^2 A\sin\frac{2\pi}{\lambda}(x - vt). \tag{5.24}$$

Similarly,

$$\frac{\partial^2 y}{\partial t^2} = -\left(\frac{2\pi v}{\lambda}\right)^2 A\sin\frac{2\pi}{\lambda}(x - vt). \tag{5.25}$$

Finally, dividing Equation (5.25) by Equation (5.24) we obtain the expected result,

$$\frac{\partial^2 y}{\partial t^2} = v^2 \frac{\partial^2 y}{\partial x^2}.$$

5.4 THE EQUATION OF A VIBRATING STRING

We now derive the equation of motion for transverse vibrations on a taut string. We will find that this is just the wave equation (5.23) and it will give us the velocity v in the latter equation in terms of the physical parameters of the system. We consider a short segment of the string and the forces that act upon it as the wave passes by. The string has mass per unit length μ and is under tension T. The wave propagates in the x-direction and the transverse displacements are in the y-direction. For small values of y the tension in the string can be assumed to be constant (cf. Section 4.6). Figure 5.8 shows the segment of the string between positions x and $x + \delta x$. Since there is a wave travelling along the string, the slopes of the string at these two positions will be different as indicated in Figure 5.8. The angles that the string makes with the x-axis are θ and $\theta + \delta\theta$ at x and $x + \delta x$, respectively. The segment of the string will be subject to a restoring force due to the tension T in the string. We can resolve this force into its components in the x- and

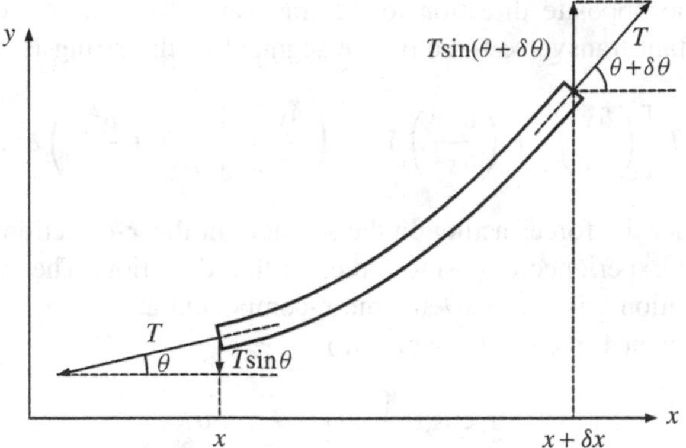

Figure 5.8 Segment of a taut string between x and $x + \delta x$, carrying a wave. The forces acting on the segment and the directions of these forces are indicated.

y-directions. We deal first with the y-component of the force, the transverse force that causes the segment to return to its equilibrium position. At x the y-component of the force F_y is equal to $T \sin\theta$. For small values of θ we have

$$\sin\theta \simeq \theta \simeq \tan\theta = \frac{\partial y}{\partial x}. \tag{5.26}$$

We see that under this condition, the transverse force at a given point is equal to the tension in the string times the slope of the string at that point, i.e.

$$F_y = T\frac{\partial y}{\partial x}. \tag{5.27}$$

Similarly, the transverse force at $x + \delta x$ is equal to the tension T times the slope at that point. The slope of the string varies smoothly and slowly from positions x to $x + \delta x$, under the assumption of small θ. Hence to a good approximation, we can say that

$$(\text{slope at } x + \delta x) = (\text{slope at } x) + (\text{rate of change of slope}) \times \delta x,$$

i.e.

$$\begin{aligned}\left(\frac{\partial y}{\partial x}\right)_{x+\delta x} &= \left(\frac{\partial y}{\partial x}\right)_x + \frac{\partial}{\partial x}\left(\frac{\partial y}{\partial x}\right)\delta x\\ &= \left(\frac{\partial y}{\partial x}\right)_x + \left(\frac{\partial^2 y}{\partial x^2}\right)\delta x.\end{aligned}$$

Hence, the transverse force at $x + \delta x$ is

$$T\left[\left(\frac{\partial y}{\partial x}\right)_x + \left(\frac{\partial^2 y}{\partial x^2}\right)\delta x\right]. \tag{5.28}$$

This acts in the opposite direction to the transverse force at x (see Figure 5.8). Thus the resultant transverse force on the segment of the string is

$$T\left[\left(\frac{\partial y}{\partial x}\right)_x + \left(\frac{\partial^2 y}{\partial x^2}\right)\delta x - \left(\frac{\partial y}{\partial x}\right)_x\right] = T\left(\frac{\partial^2 y}{\partial x^2}\right)\delta x. \tag{5.29}$$

We now consider the forces acting on the segment in the x-direction. The two ends of the segment experience opposing forces in this direction. The x-component of the force at position x is $-T\cos\theta$ and the x-component at $x + \delta x$ is $T\cos(\theta + \delta\theta)$. Hence the resultant force on the segment is

$$T\cos(\theta + \delta\theta) - T\cos\theta. \tag{5.30}$$

Since θ is small, both $\cos\theta$ and $\cos(\theta + \delta\theta)$ are both approximately equal to unity. Hence, to a good approximation, the resultant force in the x-direction is zero and there is no movement of the segment in that direction. We now use Newton's second law and Equation (5.29) to deduce the equation of motion of the segment in the y-direction. Since the mass of the segment is $\mu\delta x$, we have

$$\mu\delta x\frac{\partial^2 y}{\partial t^2} = T\left(\frac{\partial^2 y}{\partial x^2}\right)\delta x$$

or

$$\frac{\partial^2 y}{\partial t^2} = \frac{T}{\mu}\frac{\partial^2 y}{\partial x^2}. \tag{5.31}$$

This is the equation that describes wave motion on a taut string. By comparing this with the one-dimensional wave equation

$$\frac{\partial^2 y}{\partial t^2} = v^2\frac{\partial^2 y}{\partial x^2}, \tag{5.23}$$

we see that the velocity v of the wave along the string is given by

$$v = \sqrt{\frac{T}{\mu}}. \tag{5.32}$$

The velocity depends on the mass per unit length of the string and also on the tension in the string. The dimensions of $\sqrt{T/\mu}$ are [length][time]$^{-1}$ as required.

5.5 THE ENERGY IN A WAVE

In this section we turn our attention to the energy that is contained in a wave. (In Section 5.6 we will consider the rate at which this energy is transported in a travelling wave.) We again consider the case of transverse waves on a taut string and imagine the string to be divided into short segments of width δx and mass $\mu\delta x$, where μ is the mass per unit length. As the wave moves along the string, these segments will oscillate in the transverse direction and so will have kinetic energy K given by

$$K = \frac{1}{2}\mu\delta x\left(\frac{\partial y}{\partial t}\right)^2. \tag{5.33}$$

In addition, the segments will be slightly stretched when they are not at their equilibrium positions. Since the string is under tension, the segments will therefore also have potential energy U. This potential energy is equal to the extension times the tension T in the string, which we assume to be constant. To a good approximation, the extended length of a segment δs is related to the unstretched length δx (see Figure 5.9) by

$$\delta s = \frac{\delta x}{\cos\theta} = \frac{\delta x}{(1-\sin^2\theta)^{1/2}}.$$

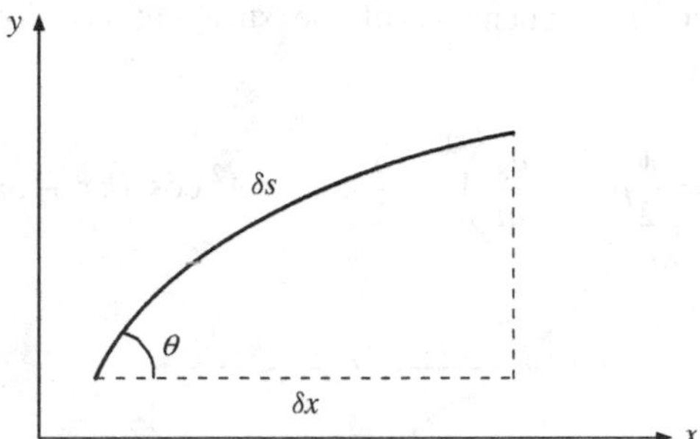

Figure 5.9 The equilibrium length δx and stretched length δs of a segment of a taut string carrying a wave.

Since θ is small,

$$\delta s \simeq \frac{\delta x}{(1-\theta^2)^{1/2}} \simeq \delta x\left(1+\frac{1}{2}\theta^2\right). \tag{5.34}$$

For small θ, we also have $\theta = \partial y/\partial x$. Thus

$$\delta s \simeq \delta x\left[1+\frac{1}{2}\left(\frac{\partial y}{\partial x}\right)^2\right]. \tag{5.35}$$

To a good approximation the potential energy is therefore given by

$$U = T(\delta s - \delta x) = \frac{1}{2}T\delta x\left(\frac{\partial y}{\partial x}\right)^2. \tag{5.36}$$

We can use Equations (5.33) and (5.36) to write down the energy in a portion $a \leq x \leq b$ of a string at time t, which is given by

$$E = \frac{1}{2}\int_a^b \mathrm{d}x \left[\mu\left(\frac{\partial y}{\partial t}\right)^2 + T\left(\frac{\partial y}{\partial x}\right)^2\right] = \frac{1}{2}\mu\int_a^b \mathrm{d}x \left[\left(\frac{\partial y}{\partial t}\right)^2 + v^2\left(\frac{\partial y}{\partial x}\right)^2\right] \tag{5.37}$$

where we have used the result $v = \sqrt{T/\mu}$, Equation (5.32). These are general results that apply to any transverse wave on the string.

As an example of the above discussion, we consider the sinusoidal wave

$$y = A\sin(kx - \omega t). \tag{5.14}$$

In particular we consider a length of the string equal to one wavelength λ. Figure 5.10(a) is a snapshot of the string between $x = x_o$ and $x = x_o + \lambda$, at a particular instant of time. It shows the variation of the instantaneous displacement y with distance x. The velocity $\partial y/\partial t = -\omega A\cos(kx - \omega t)$ and Figure 5.10(b) shows the variation of the instantaneous velocity with x. From Equation (5.33) the kinetic energy of a segment δx of the string at position x and time t is given by

$$K = \frac{1}{2}\mu\delta x\left(\frac{\partial y}{\partial t}\right)^2 = \frac{1}{2}\mu\delta x\omega^2 A^2\cos^2(kx - \omega t). \tag{5.38}$$

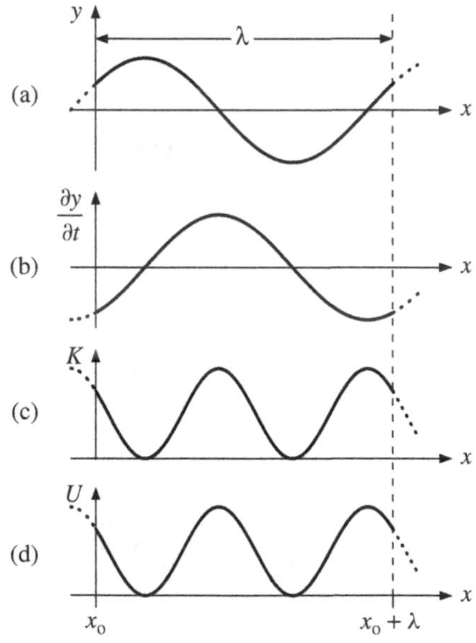

Figure 5.10 (a) Snapshot of a portion of a string carrying a travelling sinusoidal wave over one complete wavelength λ. (b) Variation of instantaneous transverse velocity of the wave $\partial y/\partial t$. (c) Variation of instantaneous kinetic energy K. (d) Variation of instantaneous potential energy U.

The resultant variation of the kinetic energy is shown in Figure 5.10(c). The total kinetic energy contained within the wavelength λ is given by

$$K_{\text{total}} = \frac{1}{2}\mu\omega^2 A^2 \int_0^{\lambda} \cos^2(kx - \omega t)\mathrm{d}x. \tag{5.39}$$

At any given instant of time, t has a fixed value and so t is a constant in the integration of Equation (5.39). Then

$$\int_0^{\lambda} \cos^2(kx - \omega t)\mathrm{d}x = \frac{\lambda}{2},$$

giving

$$K_{\text{total}} = \frac{1}{4}\mu\omega^2 A^2 \lambda. \tag{5.40}$$

The total kinetic energy in a wavelength is constant and does not change with time. [The total kinetic energy is, of course, equal to the area under the curve of Figure 5.10(c).] Similarly, we find from Equation (5.36) that the instantaneous potential energy U of a string segment at position x and time t is given by

$$\begin{aligned} U &= \frac{1}{2}v^2\mu\delta x\left(\frac{\partial y}{\partial x}\right)^2 = \frac{1}{2}v^2\mu\delta x k^2 A^2 \cos^2(kx - \omega t) \\ &= \frac{1}{2}\mu\delta x\omega^2 A^2 \cos^2(kx - \omega t) \end{aligned} \tag{5.41}$$

using $v = \omega/k$, Equation (5.15). The variation of the instantaneous potential energy with x is shown in Figure 5.10(d). The total potential energy is obtained by integrating Equation (5.41) over the complete wavelength. The result is

$$U_{\text{total}} = \frac{1}{4}\mu\omega^2 A^2 \lambda. \tag{5.42}$$

Comparing Equations (5.40) and (5.42) we see that the total kinetic energy and the total potential energy contained in a wavelength of the string are equal. The total energy in a wavelength is then given by

$$E_{\text{total}} = \frac{1}{2}\mu\omega^2 A^2 \lambda. \tag{5.43}$$

The total energy varies as the square of the amplitude of the wave and the square of the frequency of the wave. Thus the energy quadruples if we double the amplitude or double the frequency of the wave. These equations for the energies hold for all values of t.

5.6 THE TRANSPORT OF ENERGY BY A WAVE

In Section 5.5 we saw that a travelling wave contains both kinetic and potential energy and we obtained a general expression for the total energy E in a portion

$a \leq x \leq b$ of a string:

$$E = \frac{1}{2}\mu \int_a^b \mathrm{d}x \left[\left(\frac{\partial y}{\partial t} \right)^2 + v^2 \left(\frac{\partial y}{\partial x} \right)^2 \right]. \tag{5.37}$$

This equation tells us that energy is associated with a derivative of the displacement with respect to either time or position, i.e. $\partial y/\partial t$ or $\partial y/\partial x$, respectively. For a wave pulse, such as the Gaussian pulse shown in Figure 5.3, the displacement y is zero except within the finite spatial extent of the pulse. It follows from Equation (5.37) that all the energy must therefore be contained within the pulse and this energy is transported at the velocity of the pulse.

For the case of a sinusoidal wave, Figure 5.10 showed how the energy is distributed along a wavelength, at a particular instant of time. Figure 5.11 shows the displacement y and the energy distribution of part of a sinusoidal wave travelling with velocity v to the right. This figure serves to illustrate how this energy distribution is carried along with the wave at the velocity v. The total energy in a wavelength λ is given by

$$E_{\text{total}} = \frac{1}{2}\mu\omega^2 A^2 \lambda. \tag{5.43}$$

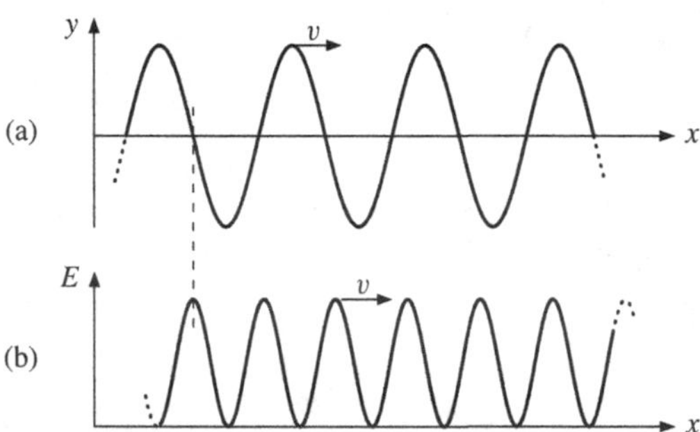

Figure 5.11 Part of a sinusoidal wave travelling at velocity v towards the right. (a) The displacement of the wave and (b) the energy distribution in the wave. The energy is carried along with the wave at velocity v.

The distance travelled by the wave in unit time is equal to v. The energy contained within this length is therefore

$$E_{\text{total}} \times \frac{v}{\lambda} = \frac{1}{2}\mu\omega^2 A^2 v.$$

This is the energy carried by the wave across any line at right angles to the direction of propagation in unit time, i.e. the power P of the wave. Hence

$$P = \frac{1}{2}\mu\omega^2 A^2 v. \tag{5.44}$$

The power of a wave depends on the square of its frequency, the square of its amplitude and its velocity.

5.7 WAVES AT DISCONTINUITIES

When a wave encounters a discontinuity at the boundary between two different media, some fraction of the wave will in general be reflected. We experience such reflections in many physical situations. If we jiggle a rope that is fixed at its other end, we observe a wave reflected travelling back towards us. We hear an echo if we clap our hands near a wall and we see that when light strikes a glass surface some of the light is reflected. In general therefore, there will be an incident wave, a transmitted wave and a reflected wave at a discontinuity. We shall now consider how the relative amplitudes and phases of these three waves can be determined. We approach this problem by considering the arrangement of two long strings smoothly joined at $x = 0$ with a constant tension along the strings. The strings have different values of mass per unit length μ, which gives rise to the discontinuity. Since the wave velocity from Equation (5.32) is given by $v = \sqrt{T/\mu}$, the wave will travel at different velocities in the two strings. The following conditions exist at the boundary between the two strings:

1. Since the two ends of the strings are joined they must move up and down together, i.e. the *displacements* of the strings at the boundary must be the same at $x = 0$ for all times. This leads to the important result that the frequency ω of the waves on both sides of the boundary must be the same. However, as the velocities of the wave are different in the two strings, the wavelengths must also be different since $\lambda = 2\pi v/\omega$ and ω is constant.
2. There must be continuity in the transverse restoring force at the boundary. Otherwise a finite difference in the force would act on an infinitesimally small mass of the string giving an infinite acceleration, which is unphysical. The transverse force is equal to $T(\partial y/\partial x)$ (cf. Section 5.4). Since the tension T is constant, the *slopes* $(\partial y/\partial x)$ of the strings on either side of the join must be the same at $x = 0$ for all times.

We now use these boundary conditions to determine the relative amplitudes and phases of the incident, transmitted and reflected waves. We let the incident wave be

$$y_I = A_1 \cos(\omega t - k_1 x), \tag{5.45}$$

where k_1 is the wavenumber in the left-hand string. We chose the cosine form so that the incident wave has its maximum value at the boundary, $x = 0$ when $t = 0$. We write the transmitted wave as

$$y_T = A_2 \cos(\omega t - k_2 x), \tag{5.46}$$

where k_2 is the wavenumber in the right-hand string and the reflected wave as

$$y_R = B_1 \cos(\omega t + k_1 x). \tag{5.47}$$

These waves are shown schematically in Figure 5.12. The resultant wave on the left-hand string y_1 is the sum of the incident and reflected waves while the resultant

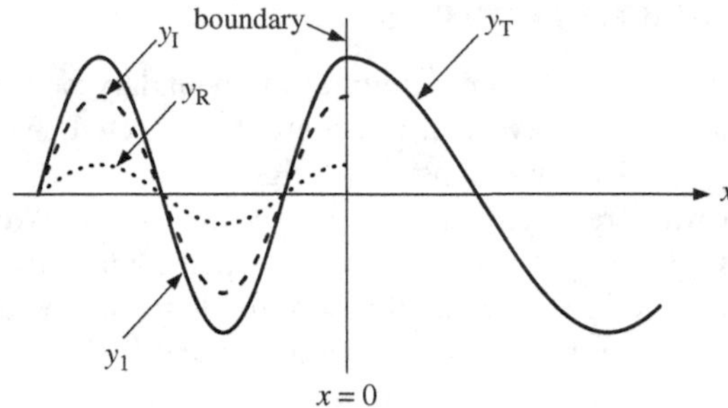

Figure 5.12 The incident, reflected and transmitted waves at the boundary of two strings of different mass per unit length. The incident wave y_I is shown as the dashed line on the left-hand side of the boundary, while the transmitted wave y_T is shown as the solid line on the right-hand side of the boundary. The reflected wave y_R is shown as the dotted line. The solid line on the left-hand side of the boundary is the sum of the incident and reflected waves, $y_1 = y_I + y_R$.

wave on the right-hand string y_2 is just the transmitted wave, i.e.,

$$y_1 = y_I + y_R \text{ and } y_2 = y_T, \tag{5.48}$$

as shown in Figure 5.12. Thus

$$y_1 = A_1 \cos(\omega t - k_1 x) + B_1 \cos(\omega t + k_1 x)$$

and

$$y_2 = A_2 \cos(\omega t - k_2 x).$$

Condition 1 gives $y_1 = y_2$ at $x = 0$. Hence

$$A_1 \cos(\omega t - k_1 x) + B_1 \cos(\omega t + k_1 x) = A_2 \cos(\omega t - k_2 x),$$

where $x = 0$. Since this equation must be true for all times we can take $t = 0$ to obtain

$$A_1 + B_1 = A_2. \tag{5.49}$$

Condition 2 gives $\partial y_1/\partial x = \partial y_2/\partial x$ at $x = 0$, for all times. Hence

$$k_1 A_1 \sin(\omega t - k_1 x) - k_1 B_1 \sin(\omega t + k_1 x) = k_2 A_2 \sin(\omega t - k_2 x),$$

where $x = 0$. This time we choose $t = \pi/2\omega$, which gives

$$k_1 A_1 - k_1 B_1 = k_2 A_2. \tag{5.50}$$

We want to find the ratio of amplitudes A_2/A_1 for the transmitted and incident waves and also the ratio B_1/A_1 for the reflected and incident waves. Manipulating Equations (5.49) and (5.50) to eliminate B_1 gives

$$\frac{A_2}{A_1} = \frac{2k_1}{k_1 + k_2} = T_{12}, \tag{5.51}$$

where T_{12} is the *transmission coefficient of amplitude*. Similarly, manipulating Equations (5.49) and (5.50) to eliminate A_2 gives

$$\frac{B_1}{A_1} = \frac{k_1 - k_2}{k_1 + k_2} = R_{12}, \tag{5.52}$$

where R_{12} is the *reflection coefficient of amplitude*. The transmission coefficient T_{12} is always a positive quantity and can have a value within the range 0 to 2. The reflection coefficient R_{12} can have both positive and negative values within the range $+1$ to -1. It also readily follows from Equations (5.51) and (5.52) that

$$T_{12} = 1 + R_{12}. \tag{5.53}$$

Equation (5.52) shows that the sign of B_1/A_1 depends on whether k_2 is less or greater than k_1. If $k_2 < k_1$, the ratio B_1/A_1 is positive and the reflected wave is in phase with the incident wave. This is the situation shown in Figure 5.12. If $k_2 > k_1$, the ratio B_1/A_1 is negative. A change of sign between B_1 and A_1 is equivalent to a phase difference of π between the reflected and incident waves. However, Equation (5.51) shows that the ratio A_2/A_1 will always be positive and so the transmitted wave will always be in phase with the incident wave.

We can see the physical meaning of Equations (5.51) and (5.52) more easily if we write them in terms of mass per unit length μ. Using Equations (5.15) and (5.32) we have $v = \omega/k = \sqrt{T/\mu}$. Since the tension T and the frequency ω of the waves are the same in both strings, the wavenumber k is directly proportional to the square root of the mass per unit length $\sqrt{\mu}$. Hence, Equation (5.51) becomes

$$\frac{A_2}{A_1} = \frac{2\sqrt{\mu_1}}{\sqrt{\mu_1} + \sqrt{\mu_2}}, \tag{5.54}$$

and Equation (5.52) becomes

$$\frac{B_1}{A_1} = \frac{\sqrt{\mu_1} - \sqrt{\mu_2}}{\sqrt{\mu_1} + \sqrt{\mu_2}}. \tag{5.55}$$

As the mass per unit length of the right-hand string increases, we have in the limit $\mu_2 \to \infty$, the situation of the wave encountering a rigid wall. In that case, Equation (5.54) shows that $A_2 = 0$ and Equation (5.55) shows that $B_1 = -A_1$. Physically this means that if the wave encounters a rigid wall, there is no transmitted wave and the wave is totally reflected with a phase change of π between the incident and reflected waves.

Worked example

Light of wavelength 584 nm in air is incident upon a block of glass of refractive index equal to 1.50. Determine (a) the velocity, (b) the frequency and (c) the wavelength of the light within the glass block.

Solution

(a) The velocity of the light in the glass v is related to the velocity of the light in air c by the refractive index n of the glass, where $n = c/v$. Hence,

$$v = \frac{3.0 \times 10^8}{1.50} = 2.0 \times 10^8 \text{ m s}^{-1}.$$

(b) The frequency of the light in the glass ν is the same as in air. Hence,

$$\nu = \frac{c}{\lambda} = \frac{3.0 \times 10^8}{584 \times 10^{-9}} = 5.14 \times 10^{14} \text{ Hz}.$$

(c) $\lambda_{\text{glass}} = \dfrac{\lambda_{\text{air}}}{n} = \dfrac{584 \times 10^{-9}}{1.50} = 389 \text{ nm}.$

Worked example

The reflection of a wave at the boundary of two strings with different values of mass per unit length can be reduced by inserting between them, a third piece of string of appropriate length and mass per unit length. Assume that the wavenumbers in the three strings are k_1, k_2 and k_3, respectively, and that $k_3 > k_2 > k_1$. Deduce an expression for the required length L of the intermediate string and find an expression for k_2 in terms of k_1 and k_3.

Solution

When a wave encounters the discontinuity at the boundary between two different strings, there will be a reflected wave. However, by inserting a third string between them, there will be two discontinuities each of which produces a reflection. By suitable choice of the length L of the intermediate string, it is possible to arrange for the two reflected waves to cancel each other by destructive interference. In Figure 5.13 the incident wave y_1, and transmitted waves y_2 and y_3 are represented. Also represented are the wave y_4 reflected at the first boundary ($x = 0$), the wave y_5 reflected at the second boundary ($x = L$) and the subsequently transmitted wave y_6. Both the reflected waves y_4 and y_5 suffer a phase change of π upon reflection since $k_3 > k_2 > k_1$. However, wave y_5 (and hence wave y_6) has to travel the additional distance $2L$ before the two waves y_4 and y_6 meet again at $x = 0$. Hence there will be a

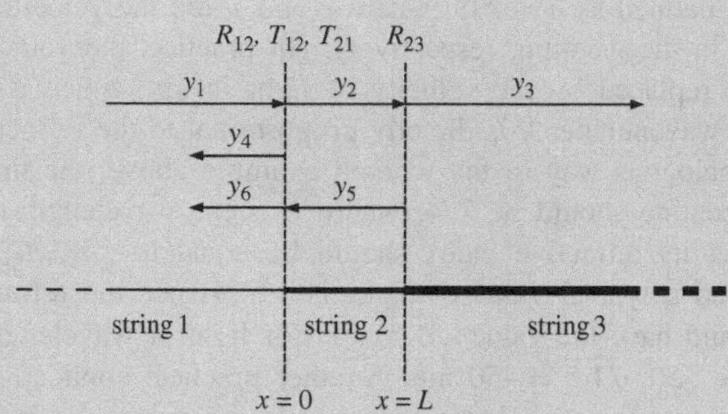

Figure 5.13 Two long strings of different mass per unit length connected by an intermediate piece of string. Also indicated are the incident, transmitted and reflected waves at the two boundaries of the three strings.

phase difference of $2\pi \times 2L/\lambda_2$ between them, where λ_2 is the wavelength in the middle string. Maximum destructive interference will occur when this phase difference is equal to π (see also Section 7.1), giving $L = \lambda_2/4$. For complete cancellation, the amplitudes of waves y_4 and y_6 should be equal. If the amplitude of incident wave y_1 is A_1, the amplitude A_4 of reflected wave y_4 will be $A_4 = R_{12}A_1$, where R_{12} is the reflection coefficient at the first boundary. The amplitude A_5 of reflected wave y_5 will be $A_5 = R_{23}A_2 = R_{23}T_{12}A_1$, where R_{23} is the reflection coefficient at the second boundary and T_{12} is the appropriate transmission coefficient at the first boundary. The amplitude A_6 of wave y_6 will be $A_6 = T_{21}A_5 = T_{21}R_{23}T_{12}A_1$, where T_{21} is the appropriate transmission coefficient at the first boundary. Hence

$$\frac{A_6}{A_4} = \frac{T_{12}R_{23}T_{21}}{R_{12}}.$$

If we make the assumption that the transmission coefficients T_{12} and T_{21} are equal to unity, which is a good assumption in many practical situations, then

$$\frac{A_6}{A_4} = \frac{R_{23}}{R_{12}}.$$

Putting $A_6 = A_4$ as required and substituting for $R_{12} = (k_1 - k_2)/(k_1 + k_2)$ and $R_{23} = (k_2 - k_3)/(k_2 + k_3)$ leads to $k_2 = \sqrt{k_1 k_3}$. (In this analysis we have neglected the contributions of waves that suffer further reflections. When all of these contributions are taken into account identical solutions are obtained.)

Analogous results of the above example have importance in many practical applications. For example, a camera lens will contain a number of different glass components and therefore many surfaces, i.e. boundaries through which the light has to pass. In order to minimise losses due to reflection, each surface is coated with a layer of appropriate thickness and refractive index. The refractive index of a medium is defined as $n = c/v$, where c and v are the velocities of the light in vacuum and in the medium, respectively. (In practice, the velocity of light in vacuum can be replaced by the velocity of light in air.) Since $v = \omega/k$ and ω is constant, the wavenumber k is directly proportional to the refractive index, i.e. $k \propto n$. In an analogous way to the worked example above, the thickness of the anti-reflection coating should be $\lambda/4$, where λ is the wavelength of the light *in the coating*, and its refractive index should be equal to $\sqrt{n_{air}n_{glass}}$. The value of n_{air} is 1.0 and a typical value of n_{glass} is 1.5. Hence the refractive index of the coating should have the value $\sqrt{1.5}$, and for light of wavelength 550 nm, its width should be $550/\sqrt{1.5} = 450$ nm. Another practical application of reducing reflection occurs in the use of ultrasonic waves to probe the human body for medical investigation. Here, the source of the ultrasonic waves is not placed in direct contact with the patient's skin. Instead a layer of suitable medium is placed between the two. The width of this intermediate layer is chosen to be equal to one-quarter of the wavelength of the ultrasonic waves in the medium and this acts to maximise the transmission of the waves into the body tissue.

5.8 WAVES IN TWO AND THREE DIMENSIONS

So far we have considered waves that propagate in one dimension. We now turn our attention to waves that propagate in two or three dimensions. An example of a two-dimensional wave is a ripple on a pond while an example of a three-dimensional wave is the sound wave produced by a fired gun. We start by considering waves on a taut membrane which is the two-dimensional analogue of the taut string. The membrane has a mass per unit area σ and is stretched uniformly under surface tension S. This tension is the force that would appear on either side of a cut in the membrane and acts in the direction at right angles to the cut. The surface tension S has units of force per unit length. Figure 5.14 shows a small element of the membrane with sides of length δx and δy. At equilibrium this element lies in the x-y plane. The forces acting at each end of

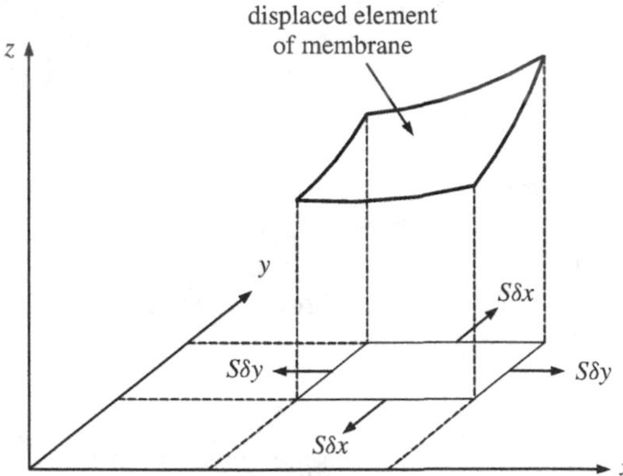

Figure 5.14 An element $\delta x \delta y$ of a taut membrane, showing the element at its equilibrium position and at a displaced position when a wave passes by.

the element are either $S\delta x$ or $S\delta y$ with directions as indicated in Figure 5.14. This figure also shows the element at some instant of time as the wave passes by. The element is displaced in the z-direction and becomes curved. As for the taut string case (Section 5.4), it is assumed that the element of the membrane only moves transversely and not sideways which is a good approximation for small displacements. We follow an analogous treatment to that of the taut string. If the displacement of the membrane element is small, the surface tension S can be assumed to be constant. Hence the magnitudes of the forces acting at each end of the element remain the same although the directions of these forces will change. This can be seen more clearly when we take a side view of the membrane element as in Figure 5.15, which shows the curvature of the element in the x-z plane. From comparison with the one-dimensional result (5.29), we see that the resultant force acting on the element in the x-z plane is given by

$$S\delta y\left[\left(\frac{\partial z}{\partial x}\right)_x + \left(\frac{\partial^2 z}{\partial x^2}\right)\delta x - \left(\frac{\partial z}{\partial x}\right)_x\right] = S\delta y\left(\frac{\partial^2 z}{\partial x^2}\right)\delta x, \tag{5.56}$$

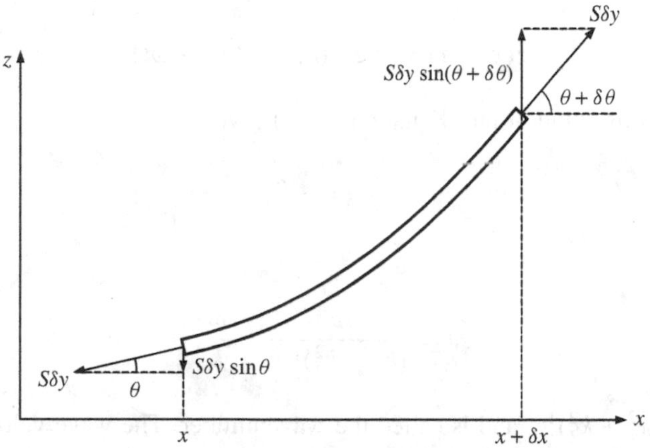

Figure 5.15 A side view, in the x-z plane, of the displaced element of the taut membrane of Figure 5.14, showing the forces acting upon it and the directions of the forces.

where we take $\theta = (\partial z/\partial x)$. This force acts in the z-direction. The membrane is also curved in the y-z plane and the resultant force due to this curvature is given by

$$S\delta x\left[\left(\frac{\partial z}{\partial y}\right)_y + \left(\frac{\partial^2 z}{\partial y^2}\right)\delta y - \left(\frac{\partial z}{\partial y}\right)_y\right] = S\delta x\left(\frac{\partial^2 z}{\partial y^2}\right)\delta y. \tag{5.57}$$

Thus the total force acting on the element in the z-direction is equal to

$$S\delta y\frac{\partial^2 z}{\partial x^2}\delta x + S\delta x\frac{\partial^2 z}{\partial y^2}\delta y.$$

Since the mass of the element is $\sigma\delta x\delta y$, we have as the equation of motion

$$\sigma\delta x\delta y\frac{\partial^2 z}{\partial t^2} = S\delta y\frac{\partial^2 z}{\partial x^2}\delta x + S\delta x\frac{\partial^2 z}{\partial y^2}\delta y, \tag{5.58}$$

giving

$$\frac{\partial^2 z}{\partial x^2} + \frac{\partial^2 z}{\partial y^2} = \frac{\sigma}{S}\frac{\partial^2 z}{\partial t^2} = \frac{1}{v^2}\frac{\partial^2 z}{\partial t^2}. \tag{5.59}$$

Equation (5.59) is the two-dimensional wave equation and we identify v as the velocity of the wave, where $v^2 = S/\sigma$.

For the case of a sinusoidal wave travelling in one dimension, we can express the wave in the form

$$y(x,t) = A\cos(kx - \omega t). \tag{5.16}$$

For a sinusoidal wave travelling in two dimensions, the corresponding solution of Equation (5.59) is

$$z(x,y,t) = A\cos(k_1 x + k_2 y - \omega t). \tag{5.60}$$

Substituting this solution into Equation (5.59) gives

$$k_1^2 + k_2^2 = \frac{\omega^2}{v^2}, \tag{5.61}$$

and hence

$$v = \frac{\omega}{(k_1^2 + k_2^2)^{1/2}} = \frac{\omega}{k}, \tag{5.62}$$

where $k = (k_1^2 + k_2^2)^{1/2}$ and is called the wavenumber. The wave velocity is equal to the angular frequency divided by the wavenumber.

We now explore the physical meaning of the solution (5.60) and the two-dimensional wave that it represents. Figure 5.16 is a pictorial representation of

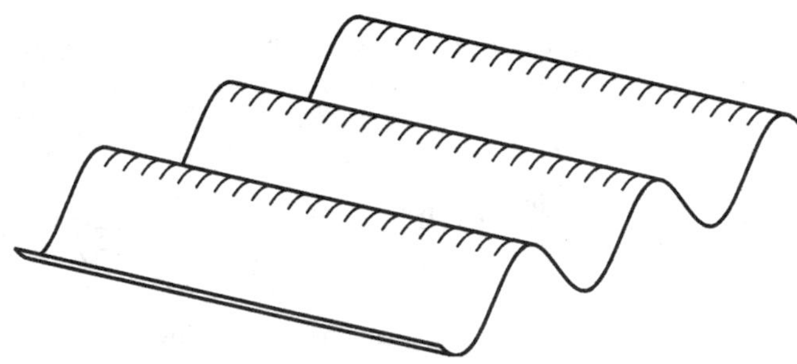

Figure 5.16 Pictorial representation of a two-dimensional wave showing the crests and troughs.

a portion of this travelling wave at some instant of time, and shows its crests and troughs. Figure 5.17 is a another view of the wave, this time from above looking down onto the x-y plane. Again it is a snapshot taken at a particular instant of time which we shall take to be $t = 0$ for convenience. The displacement z is at right angles to the x-y plane. Clearly, z varies with the independent variables x and y. However, Equation (5.60) shows that $z(x, y)_{t=0}$ will have the same value for all combinations of x and y for which $(k_1x + k_2y) = \text{constant}$. Moreover, $z(x, y)_{t=0}$ will have its maximum value when $(k_1x + k_2y) = 2\pi n$, when $n = 1, 2, 3, \ldots$. Therefore, along the x-axis (where $y = 0$) in Figure 5.17, there will be a series of maxima at $x = 2\pi n/k_1$, separated by a distance of $2\pi/k_1$. Similarly along the y-axis (where $x = 0$), there will be a series of maxima at $y = 2\pi n/k_2$, separated by a distance of $2\pi/k_2$. We can join up these sets of maxima matching the values of n as shown in Figure 5.17. Along each of these straight lines we have the condition $(k_1x + k_2y) = 2\pi n$, for the respective value of n. Hence, $z(x, y)_{t=0}$ has a constant value (and a constant phase) along each of these straight lines, which are called *wavefronts*. In this case they correspond to the maxima of the wave. (Halfway between these maxima lay the minima or troughs of the wave.) Since the wavefronts are straight, such a wave is called a *plane wave*. As time increases, these wavefronts travel at velocity v given by Equation (5.62). The direction of travel is indicated in Figure 5.17 and is at right angles to the wavefronts. We can find the direction of travel in the following way. A wavefront from Figure 5.17 is reproduced in Figure 5.18, and is denoted by the line PQ. For a wavefront we have the condition $k_1x + k_2y = 2\pi n$. We rearrange this into the form of the equation of a straight line $y = mx + c$. Then,

$$y = -\frac{k_1}{k_2}x + \frac{2\pi n}{k_2}. \tag{5.63}$$

Since $m = \tan\phi$, where ϕ is given in Figure 5.18, we have

$$\tan\phi = -\frac{k_1}{k_2}.$$

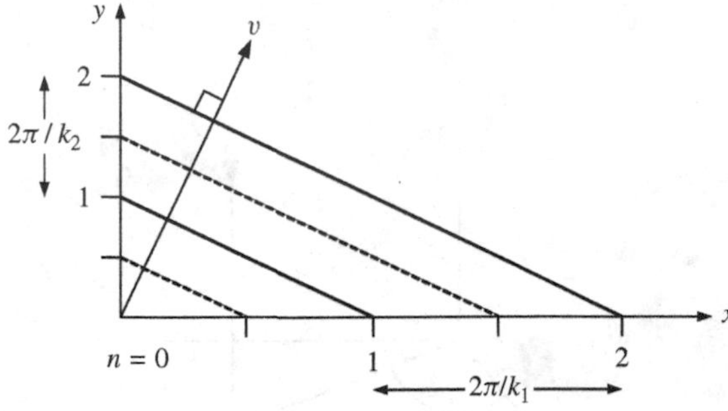

Figure 5.17 Snapshot of a two dimensional wave, looking from above onto the xy plane. The solid lines indicate the maxima (crests) of the wave while the dotted lines indicate the minima (troughs). Along these lines, which are called wavefronts, the amplitude and phase of the wave are constant. The direction of travel of the wave is at right angles to the wavefronts.

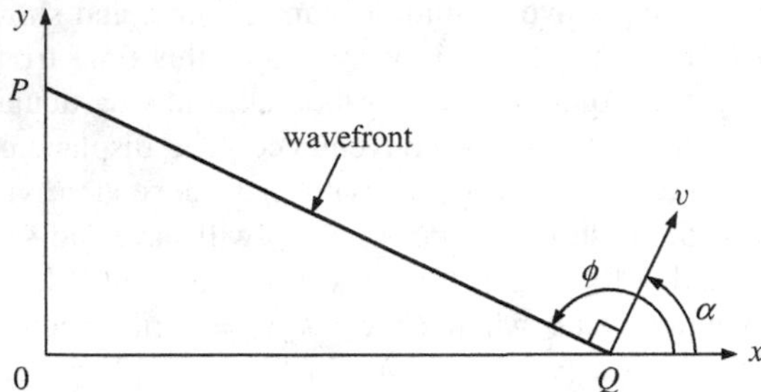

Figure 5.18 A wavefront of a two-dimensional wave showing the direction of travel of the wave characterised by the angle α.

The direction of travel, as indicated by the angle α in Figure 5.18, is at right angles to the wavefront and hence, $\alpha = (\phi - \pi/2)$. Since

$$\tan(\phi - \pi/2) = -\frac{1}{\tan\phi},$$

$$\tan\alpha = \frac{k_2}{k_1}. \tag{5.64}$$

We see that k_1 and k_2 determine the direction of travel as well as the velocity v.

5.8.1 Waves of circular or spherical symmetry

In our discussion of two-dimensional waves in Section 5.8 we defined the position of the membrane element in terms of its x- and y-coordinates. We considered the displacement z of this element in a direction at right angles to the x-y plane, describing the wave as $z = z(x, y, t)$. Moreover, we considered waves with straight wavefronts. In some situations the wavefronts are circular as in outgoing ripples on a pond. Then it is more appropriate to use the polar coordinate system illustrated in Figure 5.19. In this coordinate system a point P is specified in terms of r, θ

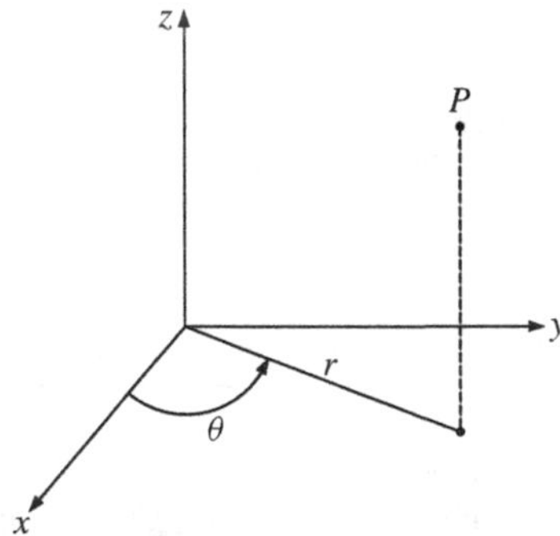

Figure 5.19 The cylindrical polar coordinate system used to describe waves with circular wavefronts. The point P is specified in terms of r, θ and z, which are independent variables.

and z, which are independent variables. Again, we specify the displacement of the element by z. For the particular case of circular waves, z has the same value for all values of θ and so we only need to consider how z depends on r and t. Hence we can express the displacement as $z = z(r, t)$. We can obtain the wave equation for circular waves from Equation (5.59) in the following way. We have $z(r) = z(x, y)$ with the condition that $r^2 = x^2 + y^2$. Then

$$\frac{\partial z}{\partial x} = \left(\frac{\partial z}{\partial r}\right)\left(\frac{\partial r}{\partial x}\right),$$

and

$$\frac{\partial^2 z}{\partial x^2} = \left(\frac{\partial^2 z}{\partial r^2}\right)\left(\frac{\partial r}{\partial x}\right)^2 + \left(\frac{\partial z}{\partial r}\right)\left(\frac{\partial^2 r}{\partial x^2}\right).$$

We have

$$\frac{\partial r}{\partial x} = \frac{x}{(x^2 + y^2)^{1/2}} = \frac{x}{r},$$

and

$$\frac{\partial^2 r}{\partial x^2} = \frac{y^2}{(x^2 + y^2)^{3/2}} = \frac{y^2}{r^3}.$$

Thus

$$\frac{\partial^2 z}{\partial x^2} = \left(\frac{\partial^2 z}{\partial r^2}\right)\left(\frac{x}{r}\right)^2 + \left(\frac{y^2}{r^3}\right)\left(\frac{\partial z}{\partial r}\right).$$

Similarly,

$$\frac{\partial^2 z}{\partial y^2} = \left(\frac{\partial^2 z}{\partial r^2}\right)\left(\frac{y}{r}\right)^2 + \left(\frac{x^2}{r^3}\right)\left(\frac{\partial z}{\partial r}\right).$$

Substituting for $\partial^2 z/\partial x^2$, $\partial^2 z/\partial y^2$ and $\partial^2 z/\partial t^2$ in Equation (5.59) we obtain

$$\frac{\partial^2 z}{\partial r^2} + \frac{1}{r}\frac{\partial z}{\partial r} = \frac{1}{v^2}\frac{\partial^2 z}{\partial t^2}. \tag{5.65}$$

This is the wave equation for two-dimensional waves of circular symmetry. Its solutions are special functions called *Bessel functions*. However, at sufficiently large values of r the second term on the left-hand side of Equation (5.65) becomes negligible compared with the first. The equation then approximates to

$$\frac{\partial^2 z}{\partial r^2} = \frac{1}{v^2}\frac{\partial^2 z}{\partial t^2}. \tag{5.66}$$

This equation has the same form as the one-dimensional wave equation and has analogous solutions such as

$$z(r,t) = A\cos(kr - \omega t), \tag{5.67}$$

where v now corresponds to the radial velocity $\mathrm{d}r/\mathrm{d}t$. Hence, circular waves emanating from a point source become plane waves at large distances from the source.

For the case of a wave propagating in a three-dimensional medium, e.g. sound waves in air, the wave equation becomes

$$\frac{\partial^2 \psi}{\partial x^2} + \frac{\partial^2 \psi}{\partial y^2} + \frac{\partial^2 \psi}{\partial z^2} = \frac{1}{v^2}\frac{\partial^2 \psi}{\partial t^2}, \tag{5.68}$$

cf. Equation (5.59). Here ψ represents the change in the relevant physical quantity that occurs as the wave passes by. For example, in the case of a sound wave ψ would correspond to changes in the local pressure of the gas. ψ is a function of the independent variables x, y, z and t, i.e. $\psi = \psi(x, y, z, t)$. Equation (5.68) has solutions such as

$$\psi = A\sin(k_1 x + k_2 y + k_3 z - \omega t), \tag{5.69}$$

where k_1, k_2 and k_3 are constants, cf. Equation (5.60), and the velocity v is given by

$$v = \frac{\omega}{(k_1^2 + k_2^2 + k_3^2)^{1/2}}, \tag{5.70}$$

cf. Equation (5.62). Again we can have situations where there is a high degree of symmetry. For example, we produce spherical outgoing sound waves when we clap our hands. For such spherical waves ψ depends only on the radial distance $r = (x^2 + y^2 + z^2)^{1/2}$ and the time t. Hence we can write $\psi = \psi(r, t)$ for which it can be shown that the wave equation (5.68) is

$$\frac{\partial^2 \psi}{\partial r^2} + \frac{2}{r}\frac{\partial \psi}{\partial r} = \frac{1}{v^2}\frac{\partial^2 \psi}{\partial t^2}. \tag{5.71}$$

To find solutions of Equation (5.71), we consider the quantity

$$u(r,t) = r\psi(r,t) \tag{5.72}$$

instead of $\psi(r, t)$. Then

$$\frac{\partial u}{\partial r} = r\frac{\partial \psi}{\partial r} + \psi, \text{ giving } \frac{\partial \psi}{\partial r} = \frac{1}{r}\left[\frac{\partial u}{\partial r} - \frac{u}{r}\right]; \tag{5.73a}$$

$$\frac{\partial^2 u}{\partial r^2} = r\frac{\partial^2 \psi}{\partial r^2} + 2\frac{\partial \psi}{\partial r}, \text{ giving } \frac{\partial^2 \psi}{\partial r^2} = \frac{1}{r}\left[\frac{\partial^2 u}{\partial r^2} - \frac{2}{r}\left(\frac{\partial u}{\partial r} - \frac{u}{r}\right)\right]; \tag{5.73b}$$

and

$$\frac{\partial^2 \psi}{\partial t^2} = \frac{1}{r}\frac{\partial^2 u}{\partial t^2}. \tag{5.73c}$$

Substituting Equation (5.73) into the wave equation (5.71) gives

$$\frac{\partial^2 u}{\partial r^2} = \frac{1}{v^2}\frac{\partial^2 u}{\partial t^2}. \tag{5.74}$$

This is the one-dimensional wave equation in the variable u. It is satisfied by solutions of the form $u = A\cos(\omega t - kr)$, giving

$$\psi = \frac{A}{r}\cos(\omega t - kr). \tag{5.75}$$

This expression for ψ shows that the amplitude of the wave (A/r) decreases as $1/r$. For a one-dimensional wave the rate of energy flowing across a line at right angles to the direction of travel is proportional to the square of the wave amplitude, cf. Equation (5.43). For a spherical wave, the energy flow crossing unit area, is again proportional to the square of the amplitude, and hence is proportional to $1/r^2$. We can see this result in a different way. As a spherical wavefront expands, the energy in the wave is spread over an increasingly large area. The area of a spherical wavefront is proportional to r^2 and hence the energy flow crossing unit area is proportional to $1/r^2$.

PROBLEMS 5

5.1 A transverse wave travelling along a string is described by the function $y = 15\cos(0.25x + 75t)$, where x and y are in millimetres and t is in seconds. Find the amplitude, wavelength, frequency and velocity of the wave. In what direction is the wave travelling?

5.2 One end of a long taut string is moved up and down in SHM with an amplitude of 0.15 m and a frequency of 10 Hz. At time $t = 0$ the end of the string (at $x = 0$) has its maximum upward displacement. The resultant wave travels down the string in the positive x-direction with a velocity of 50 m s^{-1}. Obtain an equation describing the wave.

5.3 (a) Show that the following are solutions to the one-dimensional wave equation

$$\frac{\partial^2 y}{\partial t^2} = v^2\frac{\partial^2 y}{\partial x^2}.$$

(i) $y = A\sin 2\pi\nu(t - x/v)$, (ii) $y = A\sin(2\pi/\lambda)(x + vt)$, (iii) $y = A\sin 2\pi(x/\lambda - t/T)$, (iv) $y = Ae^{i(\omega t + kx)}$, and (v) $y = A\cos(\omega_1 t - k_1 x) + B\cos(\omega_2 t - k_2 x)$, where $\omega_1/k_1 = \omega_2/k_2 = v$.

(b) Show that $\psi = A\sin(k_1 x + k_2 y + k_3 z - \omega t)$ is a solution to the three-dimensional wave equation

$$\frac{\partial^2 \psi}{\partial x^2} + \frac{\partial^2 \psi}{\partial y^2} + \frac{\partial^2 \psi}{\partial z^2} = \frac{1}{v^2}\frac{\partial^2 \psi}{\partial t^2},$$

obtaining the relationship between k_1, k_2, k_3, ω and v.

5.4 What, if any, are the differences between the waves described by:
(a) $y_1 = A\cos(\omega t - kx)$ and $y_2 = A\cos(kx - \omega t)$;
(b) $y_1 = A\sin(\omega t - kx)$ and $y_2 = A\sin(kx - \omega t)$?

5.5 A travelling wave has the profile described by $y(x,t) = A\exp[-(x - vt)^2/a^2]$, where A and a are constants. Show that the profile of the wave remains unchanged a time δt later where $v = \mathrm{d}x/\mathrm{d}t \simeq \delta x/\mathrm{d}t$. Show that this is a general result for any function of $(x - vt)$ or $(x + vt)$, i.e. $f(x - vt)$ or $g(x + vt)$.

5.6 (a) Calculate the frequencies of:
(i) a radio wave of wavelength 1500 m;
(ii) a visible photon of wavelength 500 nm;
(iii) an X-ray of wavelength 0.1 nm;
(iv) an electromagnetic wave of wavenumber 2.1 m^{-1};
(v) an ultrasonic *sound* wave that has a wavelength of 5.0 mm.
(b) Calculate the wavelengths of sound waves of frequencies 20 Hz and 15 kHz which are typical limits of a person's hearing. Compare the wavelength of a musical note of frequency 440 Hz with the typical size of a musical instrument. (Take the velocity of sound in air to be 340 m s^{-1}.)

5.7 (a) The velocity v of a wave travelling along a wire depends on the mass M of the wire and its length L and on the tension T applied to it. Use the method of dimensions to show that $v \propto \sqrt{TL/M}$. (b) Given that the six strings of a guitar have the same length and are held under similar tensions, say which of the strings will have the largest wave velocity.

5.8 (a) A horizontal wire that is 25 m long and has a mass of 100 g is held under a tension of 10 N. A sinusoidal wave of frequency 25 Hz and amplitude 15 mm travels along the wire. Calculate (i) the wave velocity along the wire and (ii) the maximum transverse velocity of any particle of the wire. (b) A square of elastic sheet of dimensions 0.75 m by 0.75 m has a mass of 125 g. A force of 2.5 N is applied to each of the four edges. What is the velocity of waves on the sheet?

5.9 A rope of length L hangs from a ceiling. (a) Show that the velocity v of a transverse wave at any point on the rope is independent of the mass and length of the rope but does depend on the distance y of the point from the bottom of the rope and that $v(y) = \sqrt{gy}$. (b) A rope of length 2.5 m hangs from a ceiling. A transverse wave is initiated at the bottom of the rope. Calculate the total time it takes for the wave to travel to the top of the rope and back to the bottom. (Assume $g = 9.81$ m s^{-2}.)

5.10 A long string is connected to an electrically driven oscillator so that a transverse sinusoidal wave is propagated along the string. The string has a mass per unit length of 30 g m^{-1} and is held under a tension of 12 N. (a) Calculate the power that must be supplied to the oscillator to sustain the propagation of the wave if it has a frequency of 150 Hz and an amplitude of 1.5 cm. (b) What will be the power required (i) if the frequency is doubled and (ii) if the amplitude is halved?

5.11 (a) A source emits waves isotropically. If the wave intensity is I_1 at a radial distance r_1 from the source, what will be the intensity I_2 at a distance r_2?
(b) The intensity of sound waves from a siren is 25 W m^{-2} at distance of 1.0 m from the siren. Assuming that the sound waves are emitted isotropically, at what distance will the sound intensity be equal to 1.0 W m^{-2} which is close to the 'threshold of pain'?

5.12 The total energy radiated by the Sun is approximately 4×10^{26} W. Estimate the solar power falling on a square metre of the Earth's surface at midday, neglecting any absorption in the atmosphere. (Take the distance from the Earth to the Sun to be 1.5×10^8 km.) What is the corresponding value at the surface of Jupiter which you can assume is five times further away from the Sun?

5.13 A long string of mass 1.0 g cm^{-1} is joined to a long string of mass 4.0 g cm^{-1} and the combination is held under constant tension. A transverse sinusoidal wave of amplitude 3.0 cm and wavelength 25 cm is launched along the lighter string. (a) Calculate the wavelength and amplitude of the wave when it is travelling along the heavier string. (b) Calculate the fraction of wave power reflected at the boundary of the two strings.

5.14 (a) Light falls normally on a glass surface. What fraction of the incident light intensity is reflected if the refractive index n of the glass is 1.5? (b) Magnesium fluoride (MgF_2) is used as an anti-reflection coating for glass lenses and has a refractive index of 1.39. What thickness of MgF_2 would be required at a wavelength λ of 550 nm for glass with $n = 1.5$? Explain why camera lenses usually have a characteristic purple colour. (c) Suppose that you wanted to *maximise* reflection at $\lambda = 550$ nm. What thickness of MgF_2 would be required for this purpose?

5.15

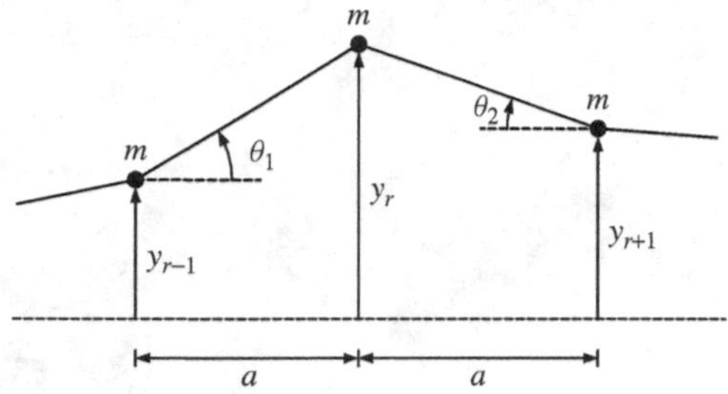

(a) The figure shows three masses that lie within a long line of identical masses of mass m connected by identical elastic strings under constant tension T with separation a. Show that the equation of motion of the central mass can be written as

$$\frac{\partial^2 y_r}{\partial t^2} = \frac{T}{m}\left[\frac{(y_{r+1} - y_r)}{a} - \frac{(y_r - y_{r-1})}{a}\right]$$

where y is the transverse displacement from equilibrium which is assumed to be small. (b) Suppose that the separation becomes infinitesimally small, $a \approx \delta x$ with $\delta x \to 0$, so that x becomes a continuous variable and the above equation can be written as

$$\frac{\partial^2 y}{\partial t^2} = \frac{T}{m}\left[\frac{y(x+\delta x) - y(x)}{\delta x} - \frac{y(x) - y(x-\delta x)}{\delta x}\right].$$

Apply a Taylor expansion to the right-hand side of this equation to show that

$$\frac{\partial^2 y}{\partial t^2} = \frac{T}{\mu}\frac{\partial^2 y}{\partial x^2}$$

where $\mu = m/\delta x$.

$$\left[\text{Taylor expansion}: y(x \pm \delta x) = y(x) \pm \delta x \frac{\partial y}{\partial x} + \frac{1}{2}(\pm\delta x)^2 \frac{\partial^2 y}{\partial x^2}.\right]$$

6

Standing Waves

In this chapter we turn our attention to *standing* waves. These are the kind of waves that occur when we pluck a guitar string. Indeed musical instruments provide a rich variety of standing waves. String instruments are plucked or bowed to set up standing waves on the strings. Blowing into the mouthpiece of a woodwind or brass instrument sets up a standing sound wave in the tubes that form the instrument. Timpani are struck to form standing waves on the drum skins. The musical instrument transforms the vibrations of the standing waves into sound waves that then propagate through the air. We will find that a standing wave can be considered as the *superposition* of two travelling waves of the same frequency and amplitude travelling in opposite directions. We will see that standing waves are the normal modes of a vibrating system and that the general motion of the system is a superposition of these normal modes. This will give us the energy of a vibrating string. It will also introduce us to the powerful technique of Fourier analysis.

6.1 STANDING WAVES ON A STRING

We shall explore the physical characteristics of standing waves by considering transverse waves on a taut string. The string is stretched between two fixed points, which we take to be at $x = 0$ and $x = L$, respectively. The transverse displacement of the string is in the y-direction. An example of such a standing wave is illustrated in Figure 6.1. Snapshots of the string at successive instants of time are shown in Figure 6.1(a)–(e), while Figure 6.1(f) shows these individual snapshots on a single set of axes. The displacement y is always zero at $x = 0$ and $x = L$ since the string is held fixed at those points. However, midway between the fixed ends we can see that the displacement of the string is also zero at all times. This point is called a *node*. Midway between this node and each end point the wave reaches its maximum displacement. These points are called *antinodes*. The positions of these

Vibrations and Waves George C. King

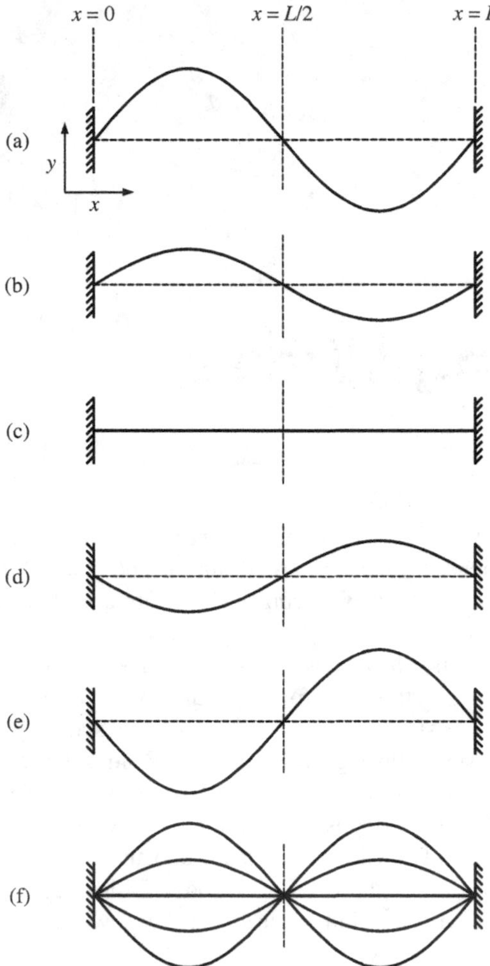

Figure 6.1 An example of a standing wave on a taut string. (a)–(e) Snapshots of the string at successive instants of time, while (f) shows these individual snapshots on a single set of axes. The displacement y is always zero at $x = 0$ and $x = L$, since the string is held fixed at those points. Midway between the fixed ends the displacement of the string is also zero at all times and this point is called a *node*. Midway between the node and each end point the wave reaches its maximum displacement and these points are called *antinodes*.

maxima and minima do not move along the x-axis with time and hence the name standing or *stationary* waves. When the string vibrates, *all* particles of the string vibrate at the same frequency. Moreover they do so in SHM about their equilibrium positions, which is the line along which the string lies when at rest. However, as shown in Figure 6.1, the amplitude of vibration of the particles varies along the length of the string. These characteristics suggest that the displacement y can be represented by

$$y(x, t) = f(x) \cos(\omega t + \phi). \tag{6.1}$$

The function $f(x)$ describes the variation of the amplitude of vibration along the x-axis. The function $\cos(\omega t + \phi)$ describes the SHM that each particle of the string undergoes. If we choose the maximum displacements of the particles to occur at $t = 0$, then the phase angle ϕ is zero and

$$y(x, t) = f(x) \cos \omega t. \tag{6.2}$$

[Imposing the condition $\phi = 0$ is equivalent to saying that initially, at $t = 0$, the string has zero velocity, i.e. from Equation (6.1)

$$\left(\frac{\partial y}{\partial t}\right)_{t=0} = -\omega f(x) \sin \phi = 0 \tag{6.3}$$

implies $\phi = 0$.] Importantly, we have written the displacement y as the product of two functions in Equation (6.2): one that depends only on x and one that depends only on t. We now substitute this solution into the one-dimensional wave equation

$$\frac{\partial^2 y}{\partial t^2} = v^2 \frac{\partial^2 y}{\partial x^2}. \tag{5.23}$$

Differentiating Equation (6.2) twice with respect to t and twice with respect to x, we obtain

$$\frac{\partial^2 y}{\partial t^2} = -\omega^2 f(x) \cos \omega t, \qquad \frac{\partial^2 y}{\partial x^2} = \frac{\partial^2 f(x)}{\partial x^2} \cos \omega t,$$

and substituting these expressions into the one-dimensional wave equation leads to

$$\frac{\partial^2 f(x)}{\partial x^2} = -\frac{\omega^2}{v^2} f(x). \tag{6.4}$$

We can compare this result with the equation of SHM:

$$\frac{d^2 x}{dt^2} = -\omega^2 x, \tag{1.6}$$

which has the general solution

$$x = A \cos \omega t + B \sin \omega t. \qquad \text{cf. (1.15)}$$

Equations (6.4) and (1.6) have the same form except the variable t in Equation (1.6) is replaced by the variable x in Equation (6.4) and x has been replaced by $f(x)$. Thus it follows that the general solution of Equation (6.4) is

$$f(x) = A \sin\left(\frac{\omega}{v} x\right) + B \cos\left(\frac{\omega}{v} x\right), \tag{6.5}$$

where A and B are constants that are determined by the *boundary conditions*. In this case the boundary conditions are $f(x) = 0$ at $x = 0$ and at $x = L$. The first condition gives $B = 0$. The second condition gives

$$A \sin\left(\frac{\omega L}{v}\right) = 0, \tag{6.6}$$

which is satisfied if

$$\frac{\omega L}{v} = n\pi, \tag{6.7}$$

where $n = 1, 2, 3, \ldots$. [Since we are not interested in the trivial solution $f(x) \equiv 0$, we exclude the value $n = 0$.] Thus, ω must take one of the values given by Equation (6.7), and so we write it as

$$\omega_n = \frac{n\pi v}{L}, \tag{6.8}$$

where for each value of n we have an associated ω_n. Substituting for $\omega = \omega_n$ in Equation (6.5) and recalling that $B = 0$, we obtain

$$f_n(x) = A_n \sin\left(\frac{n\pi}{L}x\right). \tag{6.9}$$

For each value of n we have an associated function $f_n(x)$ that is sinusoidal in shape with an associated amplitude A_n. Substituting the solution (6.9) for $f(x)$ and Equation (6.8) for $\omega = \omega_n$ in Equation (6.2) we finally obtain

$$\boxed{y_n(x, t) = A_n \sin\left(\frac{n\pi}{L}x\right) \cos \omega_n t.} \tag{6.10}$$

This equation describes the standing waves on the string, where each value of n corresponds to a different standing wave pattern. The standing wave patterns are alternatively called the *modes* of vibration of the string. As we will see in Section 6.4 these are the normal modes of the vibrating string.

The functions $f_n(x) = A_n \sin(n\pi x/L)$ for $n = 1$ to 4 are plotted in Figure 6.2(a)–(d), respectively. For the purpose of these figures the amplitudes of the four standing waves have been taken to be the same. For $n = 1$ we have

$$f_1(x) = A_1 \sin\left(\frac{\pi}{L}x\right),$$

which gives the amplitude variation shown in Figure 6.2(a). This is the fundamental mode or first *harmonic* of the string; $n = 2$ corresponds to the second harmonic, $n = 3$ corresponds to the third harmonic, etc. We see that the number of antinodes in the nth harmonic is equal to n. The corresponding angular frequencies ω_n of the standing waves are given by Equation (6.8) and are $\pi v/L$, $2\pi v/L$, $3\pi v/L$ and $4\pi v/L$, respectively. The time period T for a standing wave pattern to exactly to reproduce its shape is given by

$$T = \frac{2\pi}{\omega_n} = \frac{2L}{nv}. \tag{6.11}$$

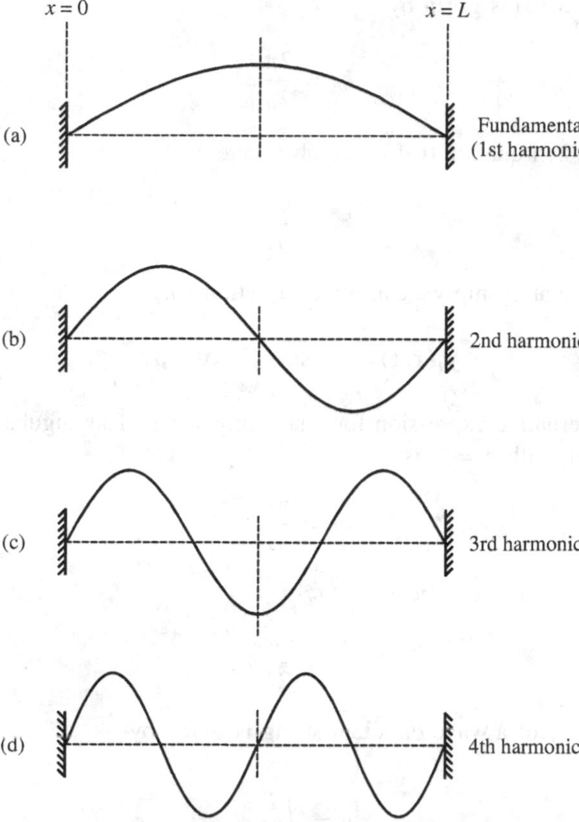

Figure 6.2 The first four harmonics for standing waves on a taut string. The first harmonic is also called the fundamental. These standing waves are described by the function $f_n(x) = A_n \sin(n\pi x/L)$ with $n = 1 - 4$. The number of antinodes in each standing wave is equal to the respective value of n.

We again define the wavelength λ of a standing wave as the repeat distance of the wave pattern. Since $v = \nu\lambda$ and $\omega = 2\pi\nu$, we can substitute for v and ω in Equation (6.11) to obtain

$$\lambda_n = \frac{2L}{n} \tag{6.12}$$

where λ_n is the wavelength of the nth standing wave. If we write this equation as

$$\frac{n\lambda_n}{2} = L \tag{6.13}$$

we see that we will obtain a standing wave only if an integral number of half-wavelengths fits between the two fixed ends of the string, as shown in Figure 6.2. Each standing wave with wavelength λ_n has a wavenumber k_n, which from Equation (5.13) is given by

$$k_n = \frac{2\pi}{\lambda_n}.$$

Since $\lambda_n = 2L/n$, Equation (6.13), we also have

$$k_n = \frac{n\pi}{L}. \tag{6.14}$$

Using this last relationship we can write Equation (6.10) as

$$y_n(x, t) = A_n \sin k_n x \cos \omega_n t, \tag{6.15}$$

which is an alternative expression for a standing wave. The angular frequency of the fundamental, with $n = 1$, is

$$\omega_1 = \frac{\pi v}{L}, \tag{6.16}$$

and its frequency, $\nu_1 = \omega_1/2\pi$, is

$$\nu_1 = \frac{v}{2L}. \tag{6.17}$$

Since the velocity of a wave on a taut string is given by

$$v = \sqrt{\frac{T}{\mu}}, \tag{5.32}$$

Equation (6.17) gives

$$\nu_1 = \frac{1}{2L}\sqrt{\frac{T}{\mu}}. \tag{6.18}$$

This equation shows how the fundamental frequency of a taut string depends on its length L, the tension T in the string and its mass per unit length μ. We can readily relate these results to stringed instruments. For example, a guitar has six strings of the same length and these are held under approximately the same tension. However, the strings have different values of mass per unit length and so their fundamental frequencies are different: the larger the mass per unit length the lower the note. Each of the strings is tuned by slightly varying the tension in the strings. The musician then plays the different notes by pressing the strings against the frets on the fingerboard to vary the length of the vibrating string. Clearly the size of a musical instrument affects the frequency or pitch of the sound it produces. This is very evident from the violin family: violin, viola, cello and double bass. These instruments steadily increase in size and produce notes of progressively lower pitch. In an analogous way the pipes of an organ steadily increase in size to produce notes of lower frequency.

As we see from Equation (6.8), the frequencies of all the harmonics of a taut string are exact multiples of the fundamental frequency and form a *harmonic series*. For most vibrating systems this is not the case. These will also vibrate at a series of higher frequencies in addition to the fundamental frequency. These higher frequencies are called *overtones*. However, in general, the frequencies of these overtones will not be an exact multiple of the fundamental: they are not harmonic. A bell, for example, will have overtones whose frequencies are not exact multiples of the fundamental. When the bell is struck, the overtone frequencies will be heard in addition to the fundamental. The skill of the bell maker is to ensure that the combination of the fundamental and the overtones produces a sound that is not discordant to the ear. (Of course, the term overtone can also be applied to a taut string but in this case the overtones are harmonic.)

We have used the example of a taut string to explore the physical characteristics of standing waves. However, standing waves occur in many different physical situations and the ideas we have been discussing are important to a wide range of physical phenomena. In a microwave oven, electromagnetic waves reflect from the walls of the oven to form standing wave patterns in the oven compartment. This means that there will inevitably be places in the compartment where the intensity of the microwave radiation is reduced and the food will not be properly cooked. To reduce the effects of these 'cold spots' the food is placed on a rotating turntable. In a laser, the light forms a standing wave between the two mirrors placed at the ends of the laser tube. In this way the wavelength of the laser light is well defined, i.e. monochromatic. In a very different example, in the realm of quantum mechanics, the discrete energy levels of atoms can be thought of as the standing-wave solutions of the Schrödinger equation.

Worked example

The Pirastro Eudoxa *A* string of a cello has a linear density $\mu = 1.70$ g m^{-1} and a length $L = 0.70$ m. The tension in the string is adjusted so that the fundamental frequency is 220 Hz. (i) What is the tension in the string? (ii) A weight of mass m is suspended from the string. What mass would produce the same tension? (iii) What is the wavelength of the sound from the string? (Take the velocity of sound in air to be 340 m s^{-1} and the acceleration due to gravity to be 9.81 m s^{-2}.)

Solution

(i) $\lambda\nu = v$ and $\lambda/2 = L$ for the fundamental frequency, giving $v = 2L\nu$.

$$T = \mu v^2 = \mu(2L\nu)^2 = \frac{1.70(2 \times 0.70 \times 220)^2}{1000} = 161 \text{ N}.$$

(ii) $m = T/g = 16.4$ kg. This result illustrates the fact that stringed instruments are subject to large forces.

(iii) The frequency of the sound wave is the same as the frequency of the vibrating string. Hence, the wavelength of the sound wave is equal to

340/220 = 1.54 m. This is different to the wavelength of the fundamental of the string ($= 2L = 1.40$ m) because of the different wave velocities in the string and in air.

Worked example

A helium-neon laser tube has a length of 0.40 m and operates at a wavelength of 633 nm. What is the difference in frequency between adjacent standing waves in the tube?

Solution

The light in a laser tube forms a standing wave between two mirrors that are placed at either end of the tube, which acts as a *resonant cavity*. Then $n\lambda/2 = L$, where n is the number of the standing wave (mode), λ is the wavelength and L is the length of the tube. Since $\lambda \ll L$, n will be very large, $\approx 1 \times 10^6$. Using $\lambda\nu = c$,

$$\nu_n = \frac{nc}{2L} \text{ and } \nu_{n+1} = \frac{(n+1)c}{2L}.$$

Hence

$$\nu_{n+1} - \nu_n = \frac{c}{2L} = \frac{3 \times 10^8}{0.80} = 3.75 \times 10^8 \text{ Hz}.$$

6.2 STANDING WAVES AS THE SUPERPOSITION OF TWO TRAVELLING WAVES

In Section 5.3, we saw that the general solution of the one-dimensional wave equation is

$$y = f(x - vt) + g(x + vt). \tag{5.4}$$

A specific example is

$$y = \frac{A}{2}\sin\frac{2\pi}{\lambda}(x - vt) + \frac{A}{2}\sin\frac{2\pi}{\lambda}(x + vt) \tag{6.19}$$

or, in terms of wavenumber $k = 2\pi/\lambda$ and angular frequency $\omega = kv$,

$$y = \frac{A}{2}\sin(kx - \omega t) + \frac{A}{2}\sin(kx + \omega t). \tag{6.20}$$

The first term in the right-hand side of this equation represents a sinusoidal wave of amplitude $A/2$ travelling in the positive x-direction and the second term represents a sinusoidal wave of amplitude $A/2$ travelling in the negative x-direction. Both waves have the same angular frequency. Using the identity

$$\sin(\alpha+\beta)+\sin(\alpha-\beta)=2\sin\alpha\cos\beta \tag{6.21}$$

we obtain

$$y=\frac{A}{2}\sin(kx-\omega t)+\frac{A}{2}\sin(kx+\omega t)=A\sin kx\cos\omega t. \tag{6.22}$$

The right-hand side of Equation (6.22) has an identical form to Equation (6.15), which we obtained for a standing wave on a taut string. Hence, we have the important result that a standing wave is the superposition of two travelling waves of the same frequency and amplitude travelling in opposite directions. This is illustrated in Figure 6.3, which shows the two travelling waves at successive instants of time separated by $T/8$ where T is the period of the wave. The wave travelling towards the right is represented by the thin continuous curve and the wave travelling towards the left is represented by the dotted curve. The arrows attached to these curves indicate the directions of travel. (At some instants of time the two waves lie on top of each other.) The thick continuous curve is the sum or superposition of the two travelling waves, i.e. the resultant standing wave. Its overall shape is just like

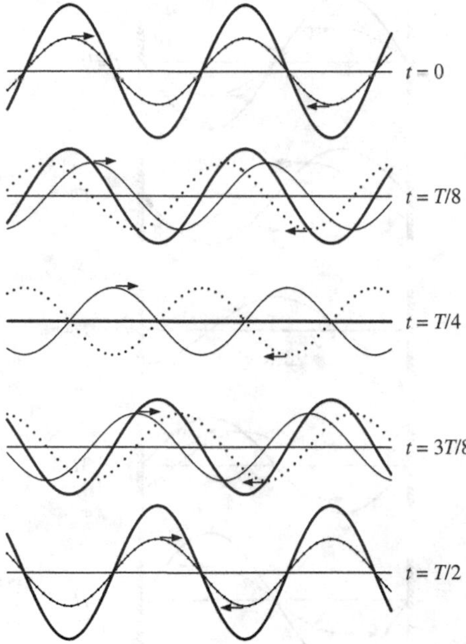

Figure 6.3 Two travelling waves of the same frequency and amplitude travelling in opposite directions, at successive instants of time. The wave travelling towards the right is represented by the thin continuous curve and the wave travelling towards the left is represented by the dotted curve. The thick continuous curve corresponds to the result of summing the two travelling waves together, i.e. the resultant standing wave. The overall shape of this curve is just like that of the standing wave corresponding to the fourth harmonic shown in Figure 6.2. As time increases, the resultant standing wave evolves as shown.

that of the standing wave corresponding to the fourth harmonic shown in Figure 6.2. As time increases the resultant standing wave evolves as shown in Figure 6.3. Any point on the standing wave is described by Equation (6.22), i.e. $y = A \sin kx \cos \omega t$. The transverse displacement of every point on the standing wave varies with SHM as $\cos \omega t$ and the amplitude of this motion varies as $A \sin kx$, i.e. the nodes and antinodes occur at fixed points on the x-axis, cf. discussion of Equation (6.1).

The two travelling sinusoidal waves that we have considered above extend to large distances in both directions (in principle to $x = \pm\infty$). A string stretched between two rigid walls has a finite length. However, it can still support standing waves. In this case it is reflections at the two walls that produce the two waves travelling in opposite directions. This is illustrated in Figure 6.4, which shows the formation of a standing wave on a string stretched between two rigid walls. The figure represents snapshots of the waves, at successive instants of time, separated by $T/8$, where T is the period of the waves. Again the thin continuous curve represents a wave travelling towards the right and the dotted curve represents a wave travelling towards the left. (At some instants of time, the incident and reflected waves lie on top of each other.) These waves are reflected at each of the walls. Inspection of Figure 6.4 shows that the waves obey the rules of reflection that we

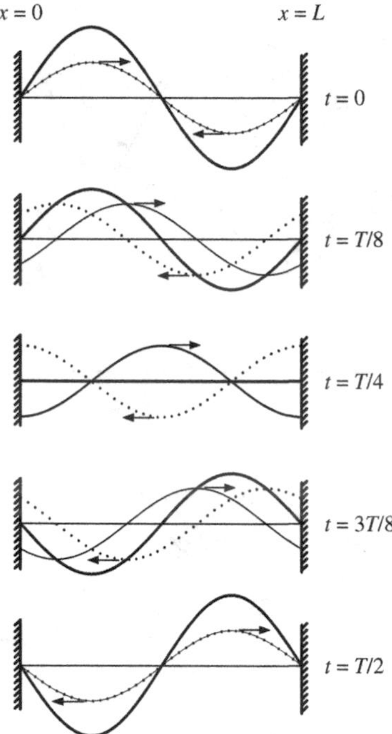

Figure 6.4 The formation of a standing wave on a string stretched between two rigid walls, at successive instants of time. The thin continuous curve represents a wave travelling towards the right and the dotted curve represents a wave travelling towards the left. These waves are reflected at each of the rigid walls. The thick continuous curve represents the result of adding the two travelling curves together, i.e. the superposition of the two waves and the resultant shape of the string.

discussed in Section 5.7 [below Equation (5.53)]: the waves reflected at a rigid boundary have the same amplitude as the incident waves but suffer a phase shift of π upon reflection. The thick continuous curve in Figure 6.4 is the superposition of the two waves and the resultant shape of the string. The formation of a standing wave and its evolution with time are apparent. Indeed the travelling waves and the resultant shape of the string shown in Figure 6.4 are identical in form to the waves shown within one wavelength on the left-hand side of Figure 6.3. We see from Figure 6.4 that the displacement of the string is always zero at the two walls, as it must be. Of course, a wave of any wavelength will be reflected at the walls. However, we can see from Figures 6.3 and 6.4 that, for a standing wave to be produced, the length of the string must be an integral multiple of half-wavelengths: $n(\lambda/2) = L$. This is just our earlier condition, Equation (6.13). If the wavelength does not meet this requirement the two travelling waves will interfere destructively and a standing wave will not result.

6.3 THE ENERGY IN A STANDING WAVE

In Section 5.5 we considered the energy of a travelling wave and found that this energy is carried along with the wave at the wave velocity. The situation for a standing wave is different. As we have seen, a standing wave is a superposition of two waves of the same frequency and amplitude travelling in *opposite* directions. The energies of these two waves are also transported in opposite directions and so there is no net transport of energy. Clearly, however, there is energy in a standing wave: a vibrating string is in motion and it stretches in moving away from its equilibrium position. Thus the string has both kinetic and potential energies. In Section 5.5 we obtained a general expression for the total energy E contained in a portion $a \leq x \leq b$ of a string that carries a transverse wave:

$$E = \frac{1}{2}\mu \int_a^b \mathrm{d}x \left[\left(\frac{\partial y}{\partial t}\right)^2 + v^2 \left(\frac{\partial y}{\partial x}\right)^2 \right], \tag{5.37}$$

where μ is the mass per unit length of the string and v is the wave velocity. The first term in the integral relates to the kinetic energy of the string and the second term to its potential energy. We now use this expression to find the total energy associated with a standing wave, i.e. the energy of a string of length L vibrating in a single mode. (The more general case where several modes are present will be considered in Section 6.4.4.) The standing wave solution for this case is given by

$$y_n(x,t) = A_n \sin\left(\frac{n\pi}{L}x\right) \cos \omega_n t, \tag{6.10}$$

where $\omega_n = v(n\pi/L)$, Equation (6.8). Differentiating this expression with respect to t and x gives

$$\begin{aligned} \frac{\partial y_n}{\partial t} &= -A_n \omega_n \sin\left(\frac{n\pi}{L}x\right) \sin \omega_n t, \\ \frac{\partial y_n}{\partial x} &= A_n \left(\frac{n\pi}{L}\right) \cos\left(\frac{n\pi}{L}x\right) \cos \omega_n t. \end{aligned} \tag{6.23}$$

Substituting the squares of these expressions into Equation (5.37), we obtain for the energy E_n of a string, of length L, vibrating in the nth mode

$$E_n = \frac{1}{2}\mu A_n^2 \left[\omega_n^2 \sin^2 \omega_n t \int_0^L \mathrm{d}x \sin^2\left(\frac{n\pi}{L}x\right) + v^2 \left(\frac{n\pi}{L}\right)^2 \cos^2 \omega_n t \int_0^L \mathrm{d}x \cos^2\left(\frac{n\pi}{L}x\right) \right]. \tag{6.24}$$

The two integrals have the same value $L/2$:

$$\int_0^L \mathrm{d}x \sin^2\left(\frac{n\pi}{L}x\right) = \int_0^L \mathrm{d}x \cos^2\left(\frac{n\pi}{L}x\right) = \frac{L}{2}. \tag{6.25}$$

To show this we use the trigonometric identity

$$\sin^2 \alpha = \frac{1}{2}[1 - \cos 2\alpha] \tag{6.26}$$

from which it follows that

$$\int_0^L \mathrm{d}x \sin^2\left(\frac{n\pi}{L}x\right) = \int_0^L \mathrm{d}x \frac{1}{2}\left[1 - \cos\left(\frac{2n\pi}{L}x\right)\right] = \frac{1}{2}\left[x - \frac{L}{2n\pi}\sin\left(\frac{2n\pi}{L}x\right)\right]_0^L = \frac{L}{2}$$

and hence

$$\int_0^L \mathrm{d}x \cos^2\left(\frac{n\pi}{L}x\right) = \int_0^L \mathrm{d}x \left[1 - \sin^2\left(\frac{n\pi}{L}x\right)\right] = L - \frac{L}{2} = \frac{L}{2}.$$

Substituting the value $L/2$ for the two integrals in Equation (6.24) and writing $v(n\pi/L) = \omega_n$, we obtain our final expression for the energy E_n of the vibrating string in the nth mode:

$$E_n = \frac{1}{4}\mu L A_n^2 \omega_n^2 (\sin^2 \omega_n t + \cos^2 \omega_n t) = \frac{1}{4}\mu L A_n^2 \omega_n^2. \tag{6.27}$$

The first term in the brackets in Equation (6.27) results from the kinetic energy of the string while the second term results from its potential energy. This equation shows that the energy of the system flows continuously between kinetic and potential energies although the total energy remains constant. This is a characteristic feature of oscillating systems, as we similarly found for the simple harmonic oscillator, Equation (1.23). When the string is at its maximum displacement, the string is instantaneously at rest and all the energy is in the form of potential energy. When the string passes through its equilibrium position, all the energy is in the form of kinetic energy. Equation (6.27) also shows that the total energy contained in the standing wave is proportional to the square of the vibration frequency and the square of the amplitude of vibration.

6.4 STANDING WAVES AS NORMAL MODES OF A VIBRATING STRING

In Chapter 4 we discussed the normal modes of a coupled oscillator. The striking characteristic of a normal mode is that all the masses move in SHM at the *same* frequency: indeed this defined the normal modes. We also saw that these normal modes are completely independent of each other and the general motion of the system is a superposition of the normal modes. All of these properties are shared by standing waves on a vibrating string; all the particles of the string perform SHM with the same frequency. Indeed the standing waves are the normal modes of the vibrating string and from now on we shall generally refer to them as normal modes. So far we have only considered the case in which a single normal mode of the string is excited. In Section 6.4.2 we shall deal with the case in which several normal modes are excited simultaneously. We shall discuss their superposition and independence and again we will see much similarity with our discussion of normal modes in Section 4.3. The methods and results that we shall demonstrate for a vibrating string admit generalisation to a huge range of physics – for example to quantum mechanics – and are therefore of great importance. We shall begin by describing the *superposition principle*.

6.4.1 The superposition principle

The superposition principle states that, if $y_1(x, t)$ and $y_2(x, t)$ are *any* two solutions of the wave equation (5.23), then so is *any* linear combination

$$y(x, t) = A_1 y_1(x, t) + A_2 y_2(x, t) \tag{6.28}$$

where A_1 and A_2 are arbitrary constants. This result follows at once from the linearity of the wave equation (5.23), i.e. each term in the wave equation is proportional to y or one of its derivatives: it does not contain quadratic or higher-power terms or product terms such as $y(\partial y/\partial x)$. (Equations of this type are known as *linear* equations.) We can see this as follows. Multiplying the first of the following equations

$$\frac{\partial^2 y_1}{\partial t^2} = v^2 \frac{\partial^2 y_1}{\partial x^2}, \qquad \frac{\partial^2 y_2}{\partial t^2} = v^2 \frac{\partial^2 y_2}{\partial x^2}$$

by A_1 and the second by A_2, and adding the resulting equations gives

$$A_1 \frac{\partial^2 y_1}{\partial t^2} + A_2 \frac{\partial^2 y_2}{\partial t^2} = v^2 \left(A_1 \frac{\partial^2 y_1}{\partial x^2} + A_2 \frac{\partial^2 y_2}{\partial x^2} \right).$$

Since

$$A_1 \frac{\partial^2 y_1}{\partial t^2} + A_2 \frac{\partial^2 y_2}{\partial t^2} = \frac{\partial^2}{\partial t^2}(A_1 y_1 + A_2 y_2), \text{ and}$$

$$v^2 \left[A_1 \frac{\partial^2 y_1}{\partial x^2} + A_2 \frac{\partial^2 y_2}{\partial x^2} \right] = v^2 \frac{\partial^2}{\partial x^2}(A_1 y_1 + A_2 y_2),$$

it follows that the linear superposition $y(x, t)$, Equation (6.28), is also a solution of the wave equation (5.23). This result clearly generalises to the superposition of any number of solutions of the wave equation. These can be *any* solutions: they do not have to be normal modes. However, for reasons that will become clearer in the course of the following discussions we now choose a general superposition of normal modes.

6.4.2 The superposition of normal modes

In Section 6.1 we found the expression for the nth normal mode of a vibrating string of length L:

$$y_n(x, t) = A_n \sin\left(\frac{n\pi}{L}x\right) \cos \omega_n t. \tag{6.10}$$

In general, the motion of the string will be a superposition of normal modes given by

$$y(x, t) = \sum_n y_n(x, t) = \sum_n A_n \sin\left(\frac{n\pi}{L}x\right) \cos \omega_n t \tag{6.29}$$

where $\omega_n = n\pi v/L$. An example of this is presented in Figure 6.5, which shows the superposition of the third normal mode with a relative amplitude of 1.0 and the thirteenth normal mode with a relative amplitude of 0.5. (We choose such a high normal mode to demonstrate the superposition of the waves more clearly.) The third normal mode is

$$y_3(x, t) = 1.0 \sin\left(\frac{3\pi}{L}x\right) \cos \omega_3 t,$$

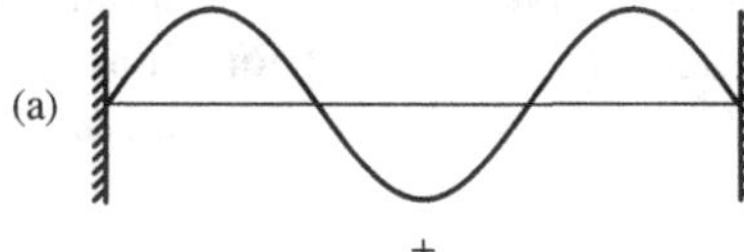

+

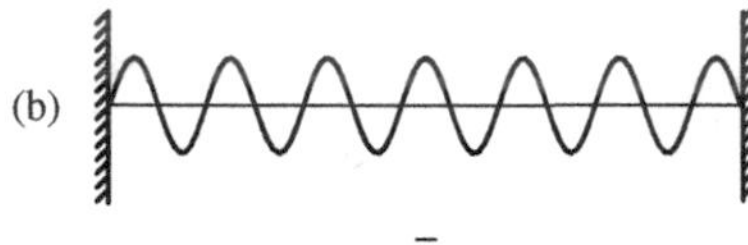

=

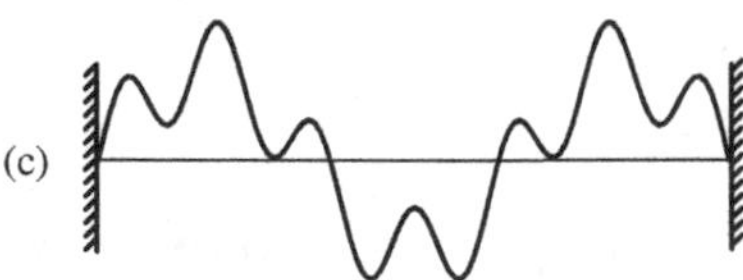

Figure 6.5 (a) Snapshot of the third harmonic $y_3(x, 0)$ of a taut string at $t = 0$. (b) Snapshot of the thirteenth harmonic $y_{13}(x, 0)$ of a taut string at $t = 0$ where the wave amplitude is equal to one half that of (a). (c) The superposition of the two harmonics to give the resultant shape of the string at $t = 0$.

and the thirteenth normal mode is

$$y_{13}(x, t) = 0.5 \sin\left(\frac{13\pi}{L}x\right) \cos \omega_{13} t.$$

Snapshots of these two normal modes at $t = 0$, i.e. $y_3(x, 0)$ and $y_{13}(x, 0)$, are shown in Figure 6.5(a) and (b), respectively. The superposition of the two normal modes is given by

$$y(x, t) = 1.0 \sin\left(\frac{3\pi}{L}x\right) \cos \omega_3 t + 0.5 \sin\left(\frac{13\pi}{L}x\right) \cos \omega_{13} t \tag{6.30}$$

and describes the motion of the vibrating string. This is illustrated in Figure 6.5(c) which again is a snapshot of the string at $t = 0$. As time increases the shape of the string evolves according to Equation (6.30). In particular it would take 13 complete periods of the higher frequency ω_{13} before the exact shape shown in Figure 6.5(c) is repeated.

To excite the two normal modes in this way, we would somehow have to constrain the shape of the string as in Figure 6.5(c) and then release it at time $t = 0$. Of course, it is impractical to do this and in practice we pluck a string to cause it to vibrate. The action of plucking a string is illustrated in Figure 6.6(a). In this example the string is displaced a distance d at one quarter of its length. Initially, the string has a triangular shape and this shape clearly does not match any of the shapes of the normal modes shown in Figure 6.2. For one thing the triangle has a sharp corner while the sinusoidal shapes of the normal modes vary smoothly.

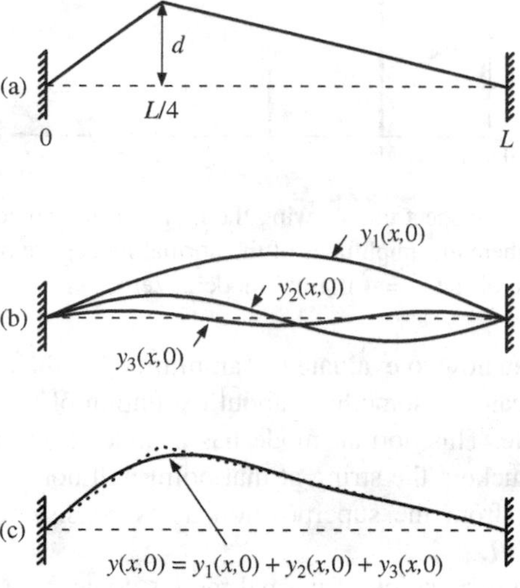

Figure 6.6 (a) The action of plucking a string is illustrated where the string is displaced a distance d at one quarter of its length. (b) The first three excited normal modes of the string. The amplitudes of these normal modes are given in the text. (c) The superposition of the first three normal modes gives a good reproduction of the initial triangular shape of the string except for the sharp corner. For all the above cases, $t = 0$.

The remarkable thing is, however, that it *is* possible to reproduce this triangular shape by adding together the normal modes of the string with appropriate amplitudes. This is illustrated by Figure 6.6. In Figure 6.6(b) the first three normal modes $y_1(x,0)$, $y_2(x,0)$ and $y_3(x,0)$ are shown. [These are given by Equation (6.10) with $t = 0$.] Their amplitudes are A, $A/2\sqrt{2}$ and $A/9$, respectively, where $A = 32d/3\pi^2$. (The general procedure for finding the values of these amplitudes is developed in Section 6.4.3.) Figure 6.6(c) shows the superposition of these three normal modes, i.e.

$$y(x,0) = y_1(x,0) + y_2(x,0) + y_3(x,0)$$

and enables a comparison with the initial shape of the string. Even using just the first three normal modes we get a surprisingly good fit to the triangular shape. By adding more normal modes, we would achieve even better agreement, especially with respect to the sharp corner. The corresponding frequencies of the normal modes are given by the usual expression $\omega_n = (n\pi v/L)$, Equation (6.8). Thus when we pluck a string we excite many of its normal modes and the subsequent motion of the string is given by the superposition of these normal modes according to Equation (6.29). A vivid way to represent the composition of the normal modes is to make a plot of their amplitudes against their frequencies which gives a *frequency spectrum*. The frequency spectrum for the example of Figure 6.6 is shown in Figure 6.7.

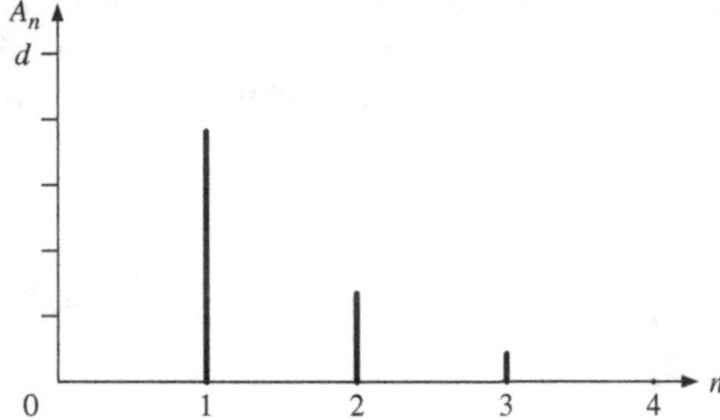

Figure 6.7 The frequency spectrum showing the first four harmonics of the plucked string shown in Figure 6.6, where the amplitudes of the normal modes are plotted against the mode number. The amplitude of the $n = 4$ normal mode is zero.

Even before we see how to evaluate the amplitudes of the excited normal modes (Section 6.4.3), we can say something about excitation of the fourth normal mode in the above example. This normal mode has a node at one quarter the length of the string. Hence, plucking the string at that point will not excite that mode which is therefore missing from the superposition as is consistent with the frequency spectrum in Figure 6.7.

Examples of the superposition of normal modes come from the sounds produced by musical instruments. The note A played on an oboe sounds distinctly different to the same note played on a flute, although both are wind instruments. In each case, the fundamental frequency or *pitch* of the note is the same. However, the relative amounts of the different normal modes (harmonics) that are produced by the two instruments are different. It is this *harmonic composition* that affects the musical quality or *timbre* of the note. The clarinet is rich in harmonics while the flute has much less harmonic content. Even different instruments of the same type may exhibit different harmonic content and so sound somewhat different. For example, the harmonic content produced by a Stradivarius violin is one of the factors that make it a very desirable instrument. We can turn this situation around and *synthesise* musical instruments. For this we use a set of oscillators to generate sinusoidal waves with the frequencies of all the harmonics we wish to include. We then add these together with appropriate relative amplitudes to synthesise the musical instrument of choice.

6.4.3 The amplitudes of normal modes and Fourier analysis

In Section 6.4.2 we saw that the general motion of a vibrating string is a superposition of normal modes, Equation (6.10). In particular, the initial shape of the string $f(x)$, i.e. at $t = 0$, is from Equation (6.29) given by

$$y(x, 0) = \sum_n A_n \sin\left(\frac{n\pi}{L}x\right) = f(x). \tag{6.31}$$

We now state a remarkable result: *any* shape $f(x)$ of the string with fixed end points $[f(0) = f(L) = 0]$ can be written as a superposition of these sine functions with *appropriate* values for the coefficients $A_1, A_2, \ldots$, i.e. in the form:

$$\boxed{f(x) = \sum_n A_n \sin\left(\frac{n\pi}{L}x\right).} \tag{6.32}$$

This result is due to Fourier. The expansion (6.32) is known as a Fourier series and the amplitudes $A_1, A_2, \ldots$ as Fourier coefficients. The idea that an essentially arbitrary function $f(x)$ can be expanded in a Fourier series can be generalised and is of great importance in much of theoretical physics and technology.

The Fourier expansion theorem, Equation (6.32), involves some difficult mathematics and we will simply assume its validity. In contrast, its application in practice is quite straightforward. Given $f(x)$, i.e. the shape of the string, the amplitudes A_n $(n = 1, 2, \ldots)$ are easily found. It is this that makes Fourier analysis such a powerful tool. The determination of the amplitudes depends on two integrals involving sine functions:

$$\int_0^L \mathrm{d}x \sin^2\left(\frac{n\pi}{L}x\right) = \frac{L}{2}, \tag{6.33}$$

$$\int_0^L \mathrm{d}x \sin\left(\frac{m\pi}{L}x\right) \sin\left(\frac{n\pi}{L}x\right) = 0, \quad m \neq n \tag{6.34}$$

where m and n are integers throughout. The first of these results we obtained earlier, Equation (6.25). For the second, we use the trignometric identity

$$\sin\alpha \sin\beta = \frac{1}{2}[\cos(\alpha - \beta) - \cos(\alpha + \beta)], \tag{6.35}$$

from which it follows that

$$\int_0^L \mathrm{d}x \sin\left(\frac{m\pi}{L}x\right)\sin\left(\frac{n\pi}{L}x\right) = \frac{1}{2}\int_0^L \mathrm{d}x\left[\cos\frac{(m-n)\pi}{L}x - \cos\frac{(m+n)\pi}{L}x\right]$$
$$= \frac{1}{2}\left[\frac{L}{(m-n)\pi}\sin\frac{(m-n)\pi}{L}x - \frac{L}{(m+n)\pi}\sin\frac{(m+n)\pi}{L}x\right]_0^L = 0,$$

for $m \neq n$, since $\sin N\pi = 0$ for $N = \pm 1, \pm 2, \ldots$.

Multiplying Equation (6.32) with $\sin(m\pi x/L)$ and integrating the resulting equation with respect to x over the range $x = 0$ to $x = L$ gives

$$\int_0^L \mathrm{d}x \sin\left(\frac{m\pi}{L}x\right) f(x) = \sum_n A_n \int_0^L \mathrm{d}x \sin\left(\frac{m\pi}{L}x\right)\sin\left(\frac{n\pi}{L}x\right). \quad (6.36)$$

It follows from Equation (6.34) that, of the terms in the series on the right-hand side of Equation (6.36), only the term with $m = n$ is different from zero, and on account of Equation (6.33) has the value $L/2$. In this way we obtain the final expression for the Fourier amplitude

$$A_n = \frac{2}{L}\int_0^L \mathrm{d}x \sin\left(\frac{n\pi}{L}x\right) f(x), \quad n = 1, 2, \ldots. \quad (6.37)$$

Equations (6.32) and (6.37) are our final result: a statement of the Fourier theorem. For any specific function $f(x)$, i.e. the shape of the string at $t = 0$, Equation (6.37) gives us the Fourier amplitudes $A_1, A_2, \ldots$. Substituting these amplitudes into Equation (6.32) gives us the initial shape of the string, expressed in its Fourier components and, from Equation (6.29), the shape of the string at subsequent times.

The situation we have described here is essentially that of classical mechanics. To solve Newton's equations of motion for a system of particles, we must specify their initial positions and velocities. For a string we have a continuum of particles, and the initial conditions become the initial position and initial velocity of each point on the string. We have treated the particular case of a string that is initially at rest, $[\partial y(x,t)/\partial t]_{t=0} = 0$, cf. Equation (6.3), and with initial shape $y(x, 0) = f(x)$. Other initial conditions are possible leading to different forms of Fourier series. We illustrate Fourier analysis by means of the following worked example.

Worked example

A string of length L is displaced at its mid-point by a distance d and released at $t = 0$. Find the first three normal modes that are excited and their amplitudes in terms of the initial displacement d.

Solution

The situation is illustrated in Figure 6.8. We represent the shape of the string at time $t = 0$ by the function $y = f(x)$. Inspection of Figure 6.8 shows that

$$f(x) = \frac{2d}{L}x, \quad 0 \le x \le L/2,$$

$$f(x) = 2d - \frac{2d}{L}x \quad L/2 \le x \le L.$$

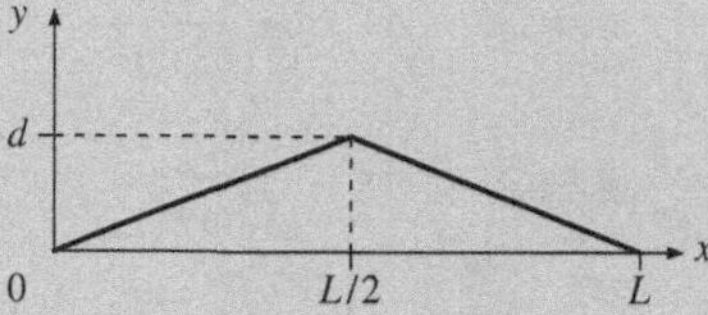

Figure 6.8 A plucked string, where its midpoint is displaced by a distance d.

To cope with the 'kink' in $f(x)$ at $x = L/2$, we split the integral (6.37) into two parts, so that

$$A_n = \frac{2}{L}\left[\int_0^{L/2} dx f(x) \sin\left(\frac{n\pi}{L}x\right) + \int_{L/2}^{L} dx f(x) \sin\left(\frac{n\pi}{L}x\right)\right].$$

Substituting for $f(x)$ over the appropriate ranges of x, the right-hand side of this equation becomes

$$\frac{2}{L}\left[\int_0^{L/2} dx \left(\frac{2d}{L}x\right) \sin\left(\frac{n\pi}{L}x\right) + \int_{L/2}^{L} dx \left(2d - \frac{2d}{L}x\right) \sin\left(\frac{n\pi}{L}x\right)\right]$$

$$= \frac{2}{L}\left[\frac{2d}{L}\int_0^{L/2} dx x \sin\left(\frac{n\pi}{L}x\right) + 2d\int_{L/2}^{L} dx \sin\left(\frac{n\pi}{L}x\right) - \frac{2d}{L}\int_{L/2}^{L} dx x \sin\left(\frac{n\pi}{L}x\right)\right].$$

We leave the evaluation of these integrals and the tidying up of the resulting expressions to the reader.[1] The final result is

$$A_n = \frac{8d}{(n\pi)^2} \sin\left(\frac{n\pi}{2}\right). \tag{6.38}$$

[1] This involves simple algebra that the reader may be inclined to follow through. The following formulae are useful for the indefinite integrals:

$$\int dx \sin ax = -\frac{1}{a}\cos ax, \quad \int dx x \sin ax = \frac{1}{a^2}\sin ax - \frac{x}{a}\cos ax,$$

where a is a constant.

It follows that when we pluck a string at its mid point we excite many normal modes (in principle an infinite number). From Equation (6.38), we have $A_n = 0$ for even values of n: we only excite those modes that have odd values of n, since modes with even n have a node at the mid-point of the string and so will not be excited. Equation (6.38) gives the amplitudes A_n of these normal modes:

$$\begin{aligned} n = 1, \quad A_1 &= \frac{8d}{(\pi)^2} \\ n = 3, \quad A_3 &= -\frac{8d}{(3\pi)^2} \\ n = 5, \quad A_5 &= \frac{8d}{(5\pi)^2} \end{aligned} \tag{6.39}$$

and the corresponding normal modes $y_n(x, t)$ are given by Equation (6.10) with these values of the amplitudes and frequencies given by $\omega_n = (n\pi/L)(\sqrt{T/\mu})$ [cf. Equations (6.8) and (5.32)]. Notice that the combination of normal modes that are excited in this example is different to that for the case of plucking the string one quarter of the way along its length, see Section (6.4.2). This has the consequence that, when plucking a violin string (playing 'pizzicato'), the timbre of the sound depends on where along the string it is plucked.

6.4.4 The energy of vibration of a string

In Section 6.3 we considered a string vibrating in a single normal mode, given by

$$y_n(x, t) = A_n \sin\left(\frac{n\pi}{L}x\right) \cos \omega_n t \tag{6.10}$$

and we derived the energy E_n of the string vibrating in this mode:

$$E_n = \frac{1}{4}\mu L A_n^2 \omega_n^2 (\sin^2 \omega_n t + \cos^2 \omega_n t) = \frac{1}{4}\mu L A_n^2 \omega_n^2. \tag{6.27}$$

We now want to obtain the energy E of the vibrating string when there are several modes present. The general superposition of normal modes is given by

$$y(x, t) = \sum_n y_n(x, t) = \sum_n A_n \sin\left(\frac{n\pi}{L}x\right) \cos \omega_n t, \tag{6.29}$$

and we must use this expression, instead of Equation (6.10), for calculating the energy E of the wave from Equation (5.37):

$$E = \frac{1}{2}\mu \int_a^b \mathrm{d}x \left[\left(\frac{\partial y}{\partial t}\right)^2 + v^2 \left(\frac{\partial y}{\partial x}\right)^2\right]. \tag{5.37}$$

The expressions for the derivatives $\partial y/\partial t$ and $\partial y/\partial x$ required in Equation (5.37) now do not consist of single terms as in Equation (6.23) for a single mode, but of sums of terms over the n modes:

$$\frac{\partial y}{\partial t} = -\sum_n A_n \omega_n \sin\left(\frac{n\pi}{L}x\right) \sin \omega_n t,$$

with a similar sum over modes for $\partial y/\partial x$. It is the squares of these derivatives that occur in Equation (5.37), and squaring these derivatives, as in

$$\left(\frac{\partial y}{\partial t}\right)^2 = \left[-\sum_m A_m \omega_m \sin\left(\frac{m\pi}{L}x\right) \cos \omega_m t\right]\left[-\sum_n A_n \omega_n \sin\left(\frac{n\pi}{L}x\right) \cos \omega_n t\right],$$

will lead to 'cross terms' containing the products

$$\sin\left(\frac{m\pi}{L}x\right) \sin\left(\frac{n\pi}{L}x\right), \quad \cos\left(\frac{m\pi}{L}x\right) \cos\left(\frac{n\pi}{L}x\right) \tag{6.40}$$

with $m \neq n$. [The cross terms containing products of cosines stem from $(\partial y/\partial x)^2$.] As a consequence, the expression for the energy E will contain integrals over these product terms, Equation (6.40), in addition to the quadratic terms which occur in Equation (6.24) for the single-mode case. However, the integrals involving the cross terms have the value 0, since for $m \neq n$

$$\int_0^L dx \sin\left(\frac{m\pi}{L}x\right) \sin\left(\frac{n\pi}{L}x\right) = \int_0^L dx \cos\left(\frac{m\pi}{L}x\right) \cos\left(\frac{n\pi}{L}x\right) = 0. \tag{6.41}$$

The first of these results was obtained in Equation (6.34), and the second is derived in exactly the same way using the trigonometric identity

$$\cos\alpha \cos\beta = \frac{1}{2}[\cos(\alpha - \beta) + \cos(\alpha + \beta)] \tag{6.42}$$

instead of Equation (6.35). Hence the cross terms with $m \neq n$ vanish in the integration and the total energy E is given by a sum of terms like Equation (6.27):

$$E = \frac{1}{4}\mu L \sum_n A_n^2 \omega_n^2 (\sin^2 \omega_n t + \cos^2 \omega_n t) = \frac{1}{4}\mu L \sum_n A_n^2 \omega_n^2. \tag{6.43}$$

The most interesting feature of this result is that each normal mode contributes an energy

$$E_n = \frac{1}{4}\mu L A_n^2 \omega_n^2 \tag{6.44}$$

quite independently of the other normal modes. This is quite typical of normal modes as we discussed in Chapter 4. They are independent of each other and there is no coupling between them. Consequently their energies are additive. [Mathematically, this independence results from Equation (6.41) which ensures that no 'cross terms' involving products of amplitudes $A_m A_n$, with $m \neq n$, survive.] An analogous result was obtained in Section 4.3 for the energy of two simple pendulums coupled by a spring. In terms of their position coordinates x_a and x_b, their motions are coupled, but in terms of their normal coordinates q_1 and q_2 they perform SHM independently of each other.

PROBLEMS 6

(Take the velocity of sound in air to be 340 m s^{-1}.)

6.1 A wire hangs vertically from a ceiling with a mass of 10 kg attached to its lower end. The wire is 0.50 m long and weighs 25 g. (a) Calculate the wave velocity along the wire and the wavelength and frequency of the fundamental mode of vibration. (b) If the maximum transverse displacement of the wire in the fundamental mode of vibration is 3.0 cm, calculate the largest values of velocity and acceleration that a particle of the wire can have.
(Assume $g = 9.81$ m s^{-2}.)

6.2 (a) A wave of frequency 262 Hz travels down a long wire that has a mass per unit length of 0.04 kg m^{-1} and a tension of 200 N. Calculate the wavelength of the wave. (b) A length L of the wire is held at a tension of 200 N between two fixed points. What value of L is required to obtain a fundamental frequency of 262 Hz (middle C) when the wire is plucked? (c) What are the frequency and wavelength of the sound wave produced by the wire when it is vibrating in its fundamental mode? Explain any differences.

6.3 (a) A taut string fastened at both ends has successive normal modes with wavelengths of 0.44 m and 0.55 m, respectively. Identify the mode numbers and determine the length of the string. (b) The cold spots in a microwave oven are found to have a separation of 0.5 cm. What is the frequency of the microwaves?

6.4 The travelling wave $y_1 = A\cos(\omega t - kx)$ combines with the reflected wave $y_2 = RA\cos(\omega t + kx)$ to produce a standing wave. Show that the standing wave can be represented by $y = 2RA\cos\omega t\cos kx + (1 - R)A\cos(\omega t - kx)$. Hence, show that the ratio of the maximum and minimum amplitudes of the standing wave is $(1 + R)/(1 - R)$.

6.5 The tension in the A string of a violin is adjusted to produce a fundamental frequency of 440 Hz. (a) What are the frequencies of the second and third harmonics? Does the wave velocity change in going to these harmonics? (b) The hearing range of the violinist extends to 15 kHz. What is the total number of harmonics of the string the violinist can hear? (c) If the violin string is 32 cm long, how far from the end of the string should the violinist place their finger to play the note of C (523 Hz)?

6.6 An octave is an increase in frequency by a factor of two. (a) Estimate the number of octaves over which you can hear. (b) Estimate the number of octaves covered by the spectrum of electromagnetic radiation from a radio frequency wave of wavelength 1500 m to a γ-ray of energy 1.0 MeV. (Planck's constant $h = 4.14 \times 10^{-15}$ eV s.)

6.7 A violin string is held under tension T. What will be the fractional change in the frequency of its fundamental mode of vibration if the tension is increased by the amount δT?

6.8 The six strings of a guitar are tuned to the notes E (lowest frequency), A, D, G, B and E (highest frequency) with a range of two octaves between the two E strings. All the strings should be held under the same tension to avoid distortion of the neck of the guitar. (a) If the high-frequency E string has a diameter of 0.30 mm, what should be the diameter of the low-frequency E string, assuming that both strings are made from the same material? (b) The fundamental frequency of the high-frequency E string is 330 Hz. If the distance between the nut and bridge of the guitar, i.e. the two fixed ends, is 65 cm and the strings are made of steel with a density of 7.7×10^3 kg m^{-3}, find the total force acting on the neck of the guitar. (c) If the high-frequency E string is made of nylon instead of steel what should be its diameter, assuming that the same tension is applied to it? (Take the density of steel to be six times the density of nylon.)

6.9 Three particles of mass $M/3$ are connected by four identical elastic strings of length $L/4$ between two rigid supports. The tension in the strings is T. (a) Show that the angular frequencies of the three normal modes for *transverse* oscillations are $\omega_1^2 = (2 - \sqrt{2})\alpha$, $\omega_2^2 = 2\alpha$ and $\omega_3^2 = (2 + \sqrt{2})\alpha$, where $\alpha = 12T/LM$. (b) Compare the frequencies $\omega_1/2\pi$, $\omega_2/2\pi$ and $\omega_3/2\pi$ with the frequencies of the first three harmonics of a string of total mass M stretched under tension T between two fixed points a distance L apart.

6.10

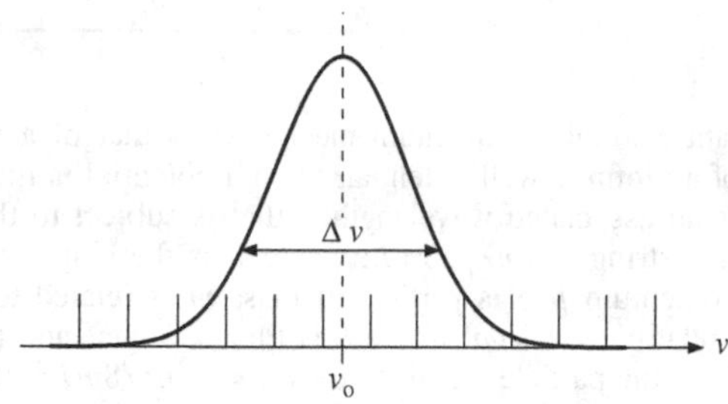

The excited atoms in the optical cavity of a laser emit light over a narrow range of frequencies and not at a single frequency. This is mainly because the atoms have a range of thermal energies and therefore a range of velocities. This is called *Doppler broadening*. The resulting spectral line profile is illustrated by the figure in which the vertical bars indicate the mode frequencies of the optical cavity. Light amplification occurs at light frequencies that coincide with a mode frequency, and that lie within a certain frequency range $\Delta\nu$, also indicated on the figure. (a) If the spectral line profile has a central frequency $\nu_o = 4.74 \times 10^{14}$ Hz and $\Delta\nu = 4.55 \times 10^9$ Hz and the length of the optical cavity is 100 cm, how many normal frequencies will occur within the range $\Delta\nu$? (b) How long would the cavity have to be so that only one mode frequency occurred within the range $\Delta\nu$?

6.11 A string is plucked one-third along its length. Give three examples of normal modes that will *not* be excited.

6.12 The function $f(x) = \alpha x$ over the range $x = 0$ to $x = L$, where α is a constant, can be represented by a Fourier series,

$$f(x) = \sum_n A_n \sin\left(\frac{n\pi}{L}x\right).$$

Show that the series is given by

$$f(x) = \frac{2\alpha L}{\pi}\left[\sin\left(\frac{\pi x}{L}\right) - \frac{1}{2}\sin\left(\frac{2\pi x}{L}\right) + \frac{1}{3}\sin\left(\frac{3\pi x}{L}\right) - \cdots\right].$$

6.13 Consider a string held under tension T between two fixed points a distance L apart. (a) If the string is displaced by a distance d at its centre show that it acquires an energy equal to $2Td^2/L$, assuming the tension in the string remains constant. (b) Using the results from the worked example in the text, show that the three harmonics of lowest frequency contain 93.3% of the energy when the string is released.

6.14 A function $f(x)$ is defined by the series

$$f(x) = \frac{4}{\pi}\left(\frac{\cos x}{1} - \frac{\cos 3x}{3} + \frac{\cos 5x}{5} - \frac{\cos 7x}{7} + \cdots\right).$$

Use a spreadsheet program to plot $f(x)$ over the range $x = 0$ to $x = 4\pi$. How would you describe the shape of the function $f(x)$?

6.15

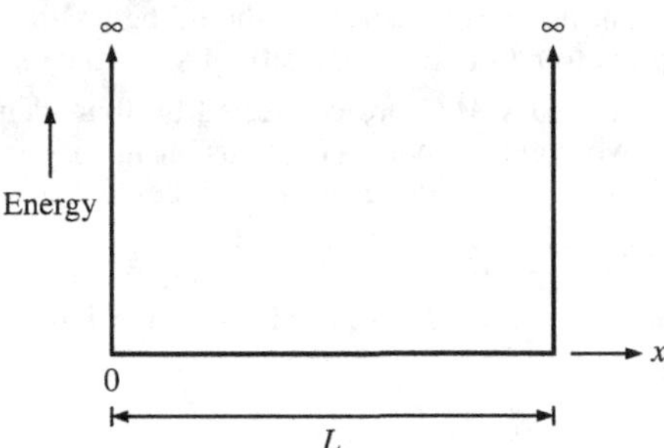

An important example in quantum mechanics is that of a particle confined between the walls of an infinite-well potential. Such a potential is illustrated in the figure. The particle has an associated wavelength λ that is subject to the same condition as that of a vibrating string, i.e. $n\lambda/2 = L$, where L is the length of the well. Moreover, the classical momentum p of a particle of mass m is related to its wavelength λ by the *de Broglie* relation $\lambda = h/p$, where h is Planck's constant. (a) Show that the allowed energies E_n of the particle are given by $E_n = n^2h^2/8mL^2$. (b) Evaluate E_n for $n = 1$, when $L = 2 \times 10^{-10}$ m, and m is the mass of the electron.
(Planck's constant $= 6.6 \times 10^{-34}$ J s and the mass of an electron $= 9.1 \times 10^{-31}$ kg.)

7

Interference and Diffraction of Waves

Interference and diffraction are some of the most striking phenomena produced by waves. Interference is evident in the rainbow of colours produced by a thin film of oil on a wet road, where the light reflected off the surface of the oil interferes with the light reflected off the water surface underneath. Diffraction is evident when water waves are incident upon the narrow mouth of a harbour. The waves spread out in a semicircular fashion after passing through the harbour mouth. We shall begin by discussing interference and later turn our attention to diffraction. However, there is no fundamental physical difference between interference and diffraction; they both result from the overlap and superposition of waves.

7.1 INTERFERENCE AND HUYGEN'S PRINCIPLE

Suppose that we have two monochromatic waves ψ_1 and ψ_2 with wavelength λ that have been derived from the same source: this avoids any random phase changes from two separate sources. These waves follow different paths and are recombined at a particular point in space. The difference in their path lengths from the common source is s. If this path difference is equal to an integral number of wavelengths, the crests and the troughs of one wave line up exactly with the crests and the troughs of the other wave, as shown in Figure 7.1(a): the two waves are said to be *in phase*. There is *constructive interference* and the amplitude of the superposition $(\psi_1 + \psi_2)$ is equal to $2A$ where A is the amplitude of the individual waves. If the path difference is an odd number of half wavelengths, the crests of one wave line up with the troughs of the other wave as shown in Figure 7.2(b): the two waves are said to be *out of phase*. There is *destructive interference* and the amplitude of their superposition is zero. We write these interference

Vibrations and Waves George C. King

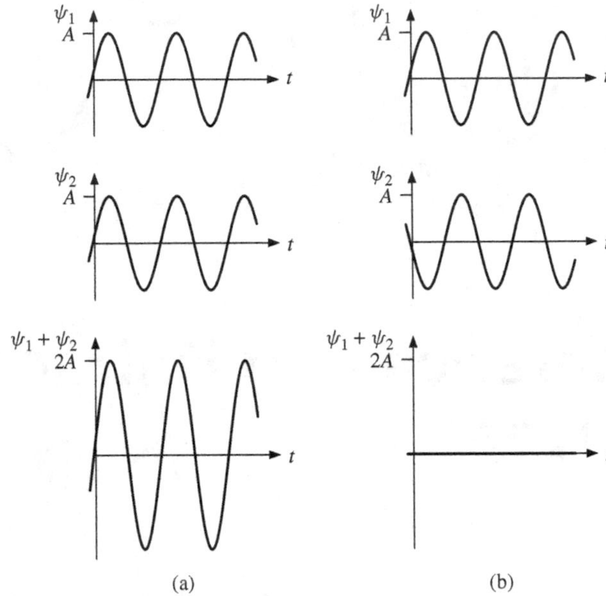

Figure 7.1 (a) Two monochromatic waves ψ_1 and ψ_2 at a particular point in space where the path difference from their common source is equal to an integral number of wavelengths. There is constructive interference and their superposition $(\psi_1 + \psi_2)$ has an amplitude that is equal to $2A$ where A is the amplitude of the individual waves. (b) The two waves ψ_1 and ψ_2 where the path difference is equal to an odd number of half wavelengths. There is destructive interference and the amplitude of their superposition is zero.

conditions as:

$$s = n\lambda, \quad n = 0, \pm 1, \pm 2, \ldots \; : \text{ constructive interference.} \tag{7.1}$$

$$s = \left(n + \frac{1}{2}\right)\lambda, \quad n = 0, \pm 1, \pm 2, \ldots \; : \text{ destructive interference.} \tag{7.2}$$

For other values of path difference s the resulting amplitude will lie between these two extremes of total constructive and destructive interference. Since phase difference $\phi = 2\pi s/\lambda$, we can also write the interference conditions as:

$$\phi = 2n\pi, \quad n = 0, \pm 1, \pm 2, \ldots \; : \text{ constructive interference.} \tag{7.3}$$

$$\phi = (2n+1)\pi, \quad n = 0, \pm 1, \pm 2, \ldots \; : \text{ destructive interference.} \tag{7.4}$$

These are the basic results for the interference of waves. They are of fundamental importance and can be applied to a wide range of physical phenomena. We shall apply them to various physical situations and in particular to an archetypal example of interference, namely Young's double-slit experiment. This experiment incorporates all the essential physical principles of wave interference and we shall discuss it in some detail. However, before doing so, we first describe *Huygen's principle*, which is named after the Dutch physicist Christian Huygen. This

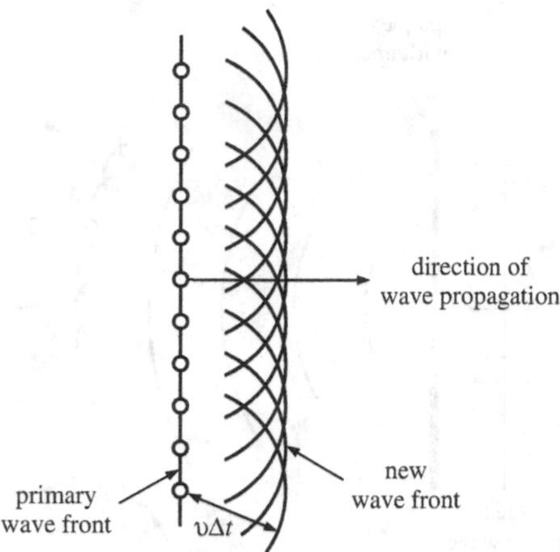

Figure 7.2 The application of Huygen's principle to the progression of a plane wave. Each point on the primary wavefront acts as a source of secondary wavelets. These secondary wavelets combine and their envelope represents the new wavefront, which is also a plane wave.

principle provides an empirical approach to predicting the progression of waves and we will use it to explain interference and, later, diffraction.

Huygen postulated that each point on a *primary* wavefront acts as a source of *secondary wavelets* such that the wavefront at some later time is the *envelope* of these wavelets. Huygen's principle is illustrated in Figure 7.2 for the example of a plane wave. To construct the wavefront at a time interval Δt later, arcs are drawn in the forward direction from points across the primary wavefront. The radius of each arc is equal to $v\Delta t$ where v is the wave velocity. These secondary wavelets combine and their envelope represents the new wavefront, which is also a plane wave as illustrated in Figure 7.2. If a wavefront encounters an aperture in an opaque barrier, the points on the wavefront across the aperture act like sources of secondary wavelets. When the aperture is very narrow, i.e. its width is comparable with the wavelength, the aperture acts like a point source and wavelets spread out in a semicircular fashion, as illustrated in Figure 7.3. The effect of the barrier is to suppress all propagation of the primary wave except through the aperture. Huygen's principle is successful in describing, at least qualitatively, the behaviour of the waves in these two examples. It is important to note, however, that Huygen's principle is an empirical approach. It provides only a qualitative description of the progression of a wave and it has shortcomings. In particular, we would expect the secondary sources on the primary wavefront to also produce a wave that propagates in the *backward* direction. In reality this does not occur and Huygen's principle ignores this other wavefront. However, a full and rigorous treatment of wave propagation, subsequently developed by G. Kirchhoff, finds that the secondary wavelets do in fact lie in the forward direction.

7.1.1 Young's double-slit experiment

Young's double-slit experiment was crucially important in confirming the wave nature of light. However, it remains of fundamental importance as an archetypal

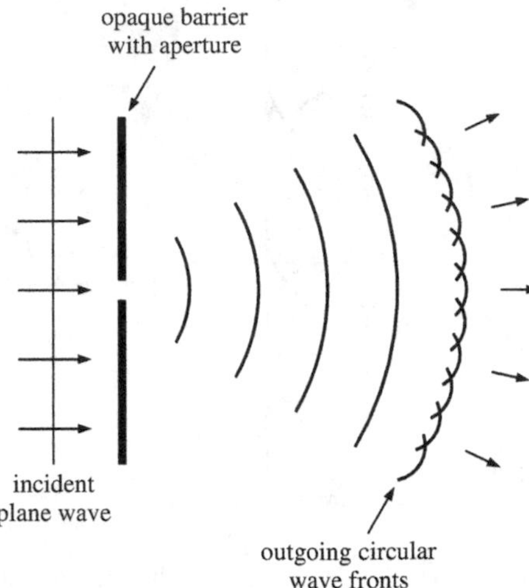

Figure 7.3 When a wavefront encounters an aperture in an opaque barrier, the barrier suppresses all propagation of the wave except through the aperture. Following Huygen's principle, the points on the wavefront across the aperture act as sources of secondary wavelets. When the width of the aperture is comparable with the wavelength, the aperture acts like a point source and the outgoing wavefronts are semicircular.

example of interference and arises, for example, in discussions of the quantum mechanical wave properties of matter. The arrangement of Young's double-slit experiment is illustrated in Figure 7.4, where the vertical scale has been greatly expanded for the sake of clarity. A monochromatic plane wave of wavelength λ is incident upon an opaque barrier that contains two very narrow slits S_1 and S_2. Each of these slits acts as a source of secondary wavelets according to Huygen's principle and the disturbance beyond the barrier is the superposition of all the wavelets spreading out from the two slits. Since these secondary wavelets are driven by the same incident wave there is a well defined phase relationship between them. This condition is called *coherence* and implies a systematic phase relationship between the secondary wavelets when they are superposed at some distant point P. It is this phase relationship that gives rise to the interference pattern, which is observed on a screen a distance L beyond the barrier. The separation of the slits is a. The slits have a long length ($\gg a$) in the direction normal to the page and this reduces the problem to two dimensions. (If we used pin holes instead of slits it would be a three-dimensional problem.) The value of a is typically $\sim$0.5 mm while the distance L to the screen is typically of the order of a few metres. Hence $L \gg a$ and this allows us to make some useful approximations as we shall see.

We consider the secondary wavelets from S_1 and S_2 arriving at an arbitrary point P on the screen. P is at a distance x from the point O that coincides with the mid-point of the two slits. The distances of S_1 and S_2 from P are l_1 and l_2, respectively. Since $L \gg a$ it can be assumed that the secondary wavelets arriving at P have the same amplitude A. The superposition of the wavelets at P gives the resultant amplitude

$$R = A[\cos(\omega t - kl_1) + \cos(\omega t - kl_2)], \tag{7.5}$$

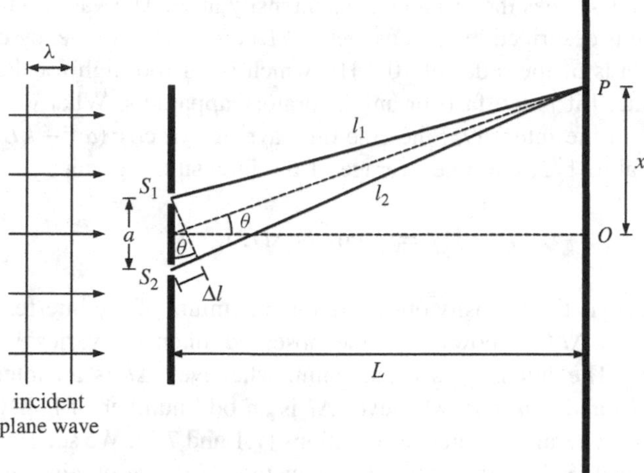

Figure 7.4 Schematic diagram of Young's double-slit experiment. The vertical scale has been enlarged for the sake of clarity. A monochromatic plane wave of wavelength λ is incident upon an opaque barrier containing two slits S_1 and S_2. These slits are very narrow but have a long length in the direction normal to the page, making this a two-dimensional problem. The resultant amplitude at point P is due to the superposition of secondary wavelets from the two slits.

where ω and k are the angular frequency and wavenumber, respectively. This result can be rewritten as

$$R = 2A\cos[\omega t - k(l_2 + l_1)/2)]\cos[k(l_2 - l_1)/2]. \tag{7.6}$$

The line joining P to the mid-point of the slits makes an angle θ with respect to the horizontal axis. Since $L \gg a$, the lines from S_1 and S_2 to P can be assumed to be parallel and also to make the same angle θ with respect to the horizontal axis. Hence

$$l_1 \simeq L/\cos\theta \simeq l_2$$

and so

$$(l_2 + l_1) \simeq 2L/\cos\theta.$$

When the two slits are separated by many wavelengths, which is the case in practice, θ is very small [cf. Equation (7.12)] and $\cos\theta \simeq 1$. Hence, we can write the resultant amplitude as

$$R = 2A\cos(\omega t - kL)\cos(k\Delta l/2) \tag{7.7}$$

where $\Delta l = (l_2 - l_1)$ is the path difference of the secondary wavelets. The intensity I at point P is equal to the square of the resultant amplitude R:

$$I = 4A^2\cos^2(\omega t - kL)\cos^2(k\Delta l/2). \tag{7.8}$$

This equation describes the instantaneous intensity at P. The variation of the intensity with time is described by the $\cos^2(\omega t - kL)$ term. The frequency of oscillation of visible light is of the order of 10^{15} Hz, which is far too high for the human eye to follow. Indeed it is too fast for any laboratory apparatus. What we observe is a *time average* of the intensity. Since the time average of $\cos^2(\omega t - kL)$ over many cycles is equal to 1/2, the time average of the intensity is given by

$$I = I_o \cos^2(k\Delta l/2), \tag{7.9}$$

where $I_o = 2A^2$ is the intensity observed at a maximum of the interference pattern. The term $\cos^2(k\Delta l/2)$ shows how the observed intensity varies with the path difference Δl. The intensity is a maximum whenever Δl is an integral number of wavelengths and it is zero whenever Δl is an odd number of half-wavelengths, illustrating the general interference conditions (7.1 and 7.2). We see from Figure 7.4 that $\Delta l \simeq a \sin\theta$. Substituting for Δl in Equation (7.9) we obtain

$$I(\theta) = I_o \cos^2(ka \sin\theta/2). \tag{7.10}$$

When θ is small so that $\sin\theta \simeq \theta$, we can write

$$\begin{aligned} I(\theta) &= I_o \cos^2(ka\theta/2) \\ &= I_o \cos^2(\pi a\theta/\lambda) \end{aligned} \tag{7.11}$$

using $k = 2\pi/\lambda$. A plot of $I(\theta)$ against θ is shown in Figure 7.5. We see that the resulting interference pattern on the screen consists of alternate bright and dark *interference fringes*.

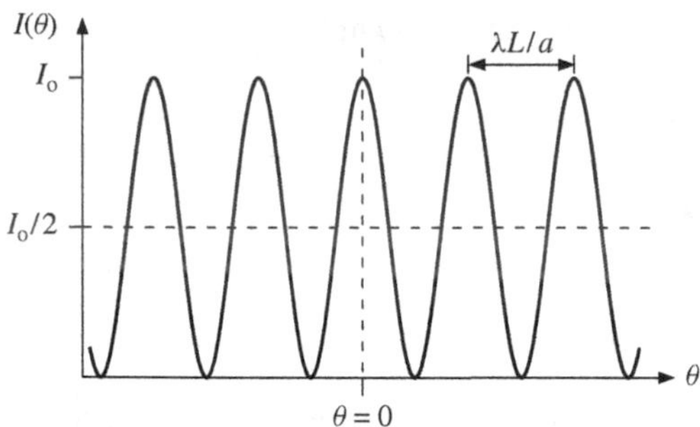

Figure 7.5 The interference pattern observed in Young's double-slit experiment. The light intensity $I(\theta)$ is plotted as a function of the angle θ shown in Figure 7.4. The small angle approximation, $\sin\theta \simeq \theta$, has been made and the separation of the bright fringes is equal to $\lambda L/a$. If there were no interference, the intensity would be uniform and equal to $I_o/2$ as indicated by the horizontal dashed line.

The important parameter that determines the general appearance of the interference pattern is the dimensionless ratio of the slit separation a to the wavelength λ.

Intensity maxima occur when

$$\theta = \frac{n\lambda}{a}, \quad n = 0, \pm 1, \pm 2, \ldots, \tag{7.12}$$

and so the bright fringes occur at distances from the point O given by

$$x = L\theta = n\frac{\lambda L}{a}, \quad n = 0, \pm 1, \pm 2, \ldots. \tag{7.13}$$

Similarly, minima occur when

$$x = \left(n + \frac{1}{2}\right)\frac{\lambda L}{a}, \quad n = 0, \pm 1, \pm 2, \ldots. \tag{7.14}$$

The distance between adjacent bright fringes is

$$x_{n+1} - x_n = \frac{\lambda L}{a} \tag{7.15}$$

and is independent of the value of n. For example, for values of $\lambda = 550$ nm, $L = 2.0$ m and $a = 0.5$ mm, the fringe separation is 2.2 mm. If there were more than a single wavelength in the incident light beam, each wavelength component would give rise to a set of bright and dark fringes. However, these would occur at different positions to those of the other wavelength components and this would cause the interference pattern to become washed out. Consequently we must use monochromatic light to obtain a clear set of interference fringes with high *visibility*.

We emphasise that there would be no interference pattern if the two sources of secondary wavelets S_1 and S_2 were not coherent. Instead the resultant intensity would be uniform across the screen with a value equal to $I_o/2$, as indicated by the horizontal dashed line in Figure 7.5. Of course, energy must be conserved, and when we have interference there is a redistribution of intensity from the regions of destructive interference to those of constructive interference. We also note that the phase difference of the secondary wavelets arriving at a point P is much more sensitive to path difference Δl than is their amplitudes. A change in Δl of $\lambda/2$ can cause the resultant intensity to go from maximum to minimum, while the wave amplitudes ($\propto 1/l$, for a two-dimensional wave) would change by a negligible amount.

We could ensure that the secondary wavelets from the two slits S_1 and S_2 are coherent, i.e. have a well defined phase relationship, by illuminating them with a point source. In practice, however, real sources are not ideal point sources because they have a finite width. Such real sources will, in general, consist of many individual point sources spread across this finite width. Moreover, these individual point sources are *not coherent* with each other.[1] For example, the source could be a slit in the jacket surrounding a sodium discharge lamp. The light from such a lamp comes from excited atoms that decay randomly and *independently* and therefore act as individual point sources that are not coherent with each other. However,

[1] This discussion relates to conventional light sources like sodium lamps and not to lasers, which are essentially coherent across the width of the light beam.

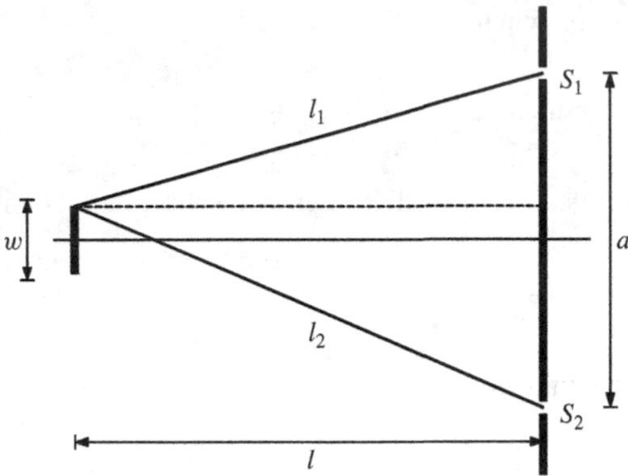

Figure 7.6 An extended source of width w that is used to illuminate the two slits in a Young's double-slit experiment.

we can still obtain an interference pattern with such a source if its spatial extent is smaller than a critical value, as we shall now show. Figure 7.6 shows an extended source of width w that is used to illuminate the two slits S_1 and S_2. The slits have a separation a and the source is at a distance l from the opaque barrier containing the slits. We consider the extended source to be made up of independent point sources that are not coherent with each other. Each of these *individual* point sources will produce secondary wavelets at S_1 and S_2 that have a well defined phase relationship. Hence, these wavelets will produce an interference pattern on a screen placed beyond the slits. However, the interference patterns produced by different point sources will be displaced relative to each other by an amount that depends on their position in the extended source. This is because the phase relationship between the secondary wavelets at S_1 and S_2 due to a particular point source depends on the path difference between that source and the two slits. In turn, the position of say a maximum in the interference pattern depends on the phase between these secondary wavelets. Clearly, if the range of phase differences between secondary wavelets at S_1 and S_2 arising from different point sources is too large, the interference pattern will become washed out. The smallest path difference is zero, which results from a point source at the centre of the extended source. (In that case the phase difference between the wavelets at S_1 and S_2 is zero.) The largest path difference will be for a point source at the end of the extended source, as illustrated in Figure 7.6, where the respective path lengths are l_1 and l_2. We have

$$l_1^2 = l^2 + (a/2 - w/2)^2, \quad l_2^2 = l^2 + (a/2 + w/2)^2,$$

giving,

$$l_2^2 - l_1^2 = aw.$$

Since $l \gg a$ and $l \gg w$,

$$l_2^2 - l_1^2 = (l_2 - l_1)(l_2 + l_1) \simeq 2l(l_2 - l_1).$$

Hence,

$$(l_2 - l_1) \simeq \frac{aw}{2l}. \tag{7.16}$$

To obtain a clear interference pattern the range of the phase differences for the wavelets produced at S_1 and S_2 must be sufficiently small. In terms of path difference, this means that $(l_2 - l_1)$ must be much less than the wavelength λ, i.e.

$$\frac{aw}{2l} \ll \lambda \tag{7.17}$$

and hence,

$$w \ll \frac{2l\lambda}{a}. \tag{7.18}$$

Thus an extended source of width w behaves like a coherent light source so long as Equation (7.18) is satisfied. The extended source subtends an angle θ at each slit where

$$\theta \simeq \frac{w}{l}. \tag{7.19}$$

Thus, from Equation (7.18), we have

$$\theta \ll \frac{2\lambda}{a} \tag{7.20}$$

which gives the maximum divergence angle that the source can have to produce clear interference fringes. If, for example, $a = 0.5$ mm, then θ must be much less than 10^{-3} rad at a wavelength of 500 nm. Hence, if we used a discharge lamp that operated at this wavelength and it was placed a distance of 1 m from the two slits, we would have to place the lamp behind a slit of width less than 1 mm. These consideration apply more generally to systems containing many slits. For the example of a diffraction grating, a is the distance between the outermost slits, i.e. the size of the diffraction grating. Hence, Equation (7.20) relates the size of the diffraction grating to the angle subtended at the grating by the extended source.

Interference occurs in many other physical situations as, for example, with sound waves. This is illustrated in Figure 7.7, which shows two loudspeakers that are connected to the *same* amplifier. Since the loudspeakers are driven by the same amplifier, the sound waves are coherent and will produce an interference pattern. The resulting sound intensity is plotted as a function of distance along the line AB which is at a large distance from the loudspeakers compared to their separation. If we were to move along that line we would hear the sound intensity rise and fall. In contrast, there would no interference if the loudspeakers were driven by different amplifiers, since there would be nothing to maintain a constant phase relationship between the sound waves. This means we would not experience interference effects in front of the stage at a rock concert if there were two guitarists using separate

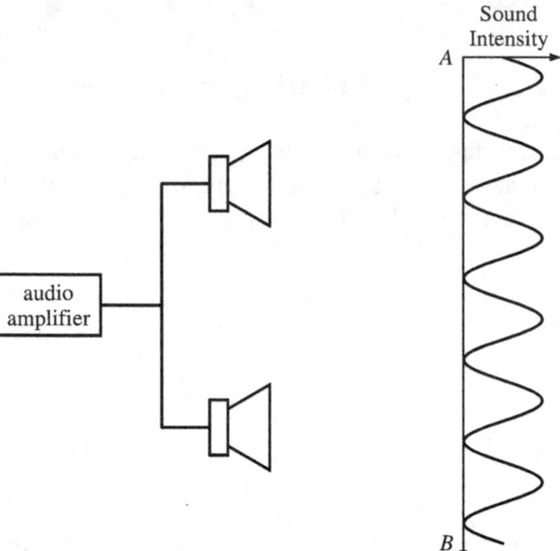

Figure 7.7 (Not to scale.) Two loudspeakers connected to the same amplifier produce coherent sound waves. These waves superpose to produce an interference pattern along the line *AB*. The intensity of the sound is proportional to the square of the amplitude of the superposition and the variation in intensity along the line *AB* is shown. This line is at a large distance from the loudspeakers compared to their separation.

amplifiers even if they were playing the same note. Interference is also exploited in a range of practical applications. For example, when a beam of X-rays is shone onto a crystal it is found that the intensity of the reflected rays becomes intense at certain values of the angle θ that the incident beam makes with the atomic planes of the crystal. This occurs because the X-rays are reflected off successive atomic planes and if the resultant path difference is equal to an integral number of wavelengths, there is constructive interference. The angles for constructive interference are given by the *Bragg law:*

$$2d \sin\theta = n\lambda, \quad n = \pm 1, \pm 2, \ldots$$

where d is the separation of the atomic planes and λ is the wavelength. X-ray crystallography is widely used to determine the structure of matter and Crick and Watson famously got their idea for the double-helix structure of DNA from looking at Rosalind Franklin's X-ray interference patterns from DNA.

7.1.2 Michelson spectral interferometer

Young's double-slit experiment is an example of interference by *division of wavefront*, where we take two portions of the wavefront to obtain the two coherent wave sources. We can also have interference by *division of amplitude* where the primary wave itself is divided into two parts by, for example, a semi-silvered mirror. An important example of division of amplitude is the *Michelson spectral interferometer*. This interferometer provided one of the key experimental observations underpinning the theory of relativity. It is also a powerful research tool with many applications including the determination of the emission spectra of atoms and molecules, i.e. the wavelengths of their emitted radiations. In particular, it can do this with very high spectral resolution. The principle of operation of the Michelson spectral interferometer is illustrated in Figure 7.8. A beam of light from

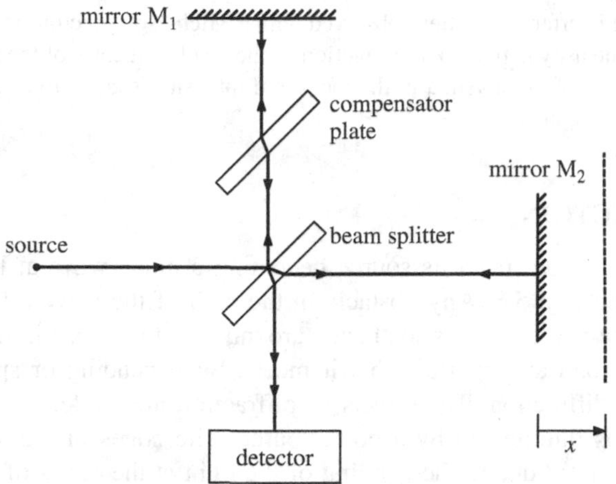

Figure 7.8 Schematic diagram of the Michelson spectral interferometer.

a monochromatic source is split into two equal beams by the semi-reflecting front face of the *beam splitter*. The two separate beams travel to mirrors M_1 and M_2, respectively, and then return to the beamsplitter from where they travel along the same path to the detector. The presence of the *compensator plate* ensures that the beams transverse the same total thickness of glass in both arms of the interferometer. The two superposed beams have the same intensity at the detector since each undergoes one transmission and one reflection at the semi-reflecting surface of the beamsplitter. Mirror M_1 is fixed in position. The position of mirror M_2 can be adjusted with a very fine micrometer screw. If the path lengths of the two beams are the same or are different by an integral number of wavelengths, the beams will interfere constructively at the detector and there will be a maximum in the detected light intensity. However, if the path lengths are different by an odd number of half-wavelengths, there will be destructive interference and the detected light intensity will be zero. When the detected light intensity is plotted as a function of the displacement x of mirror M_2 an interference pattern is obtained, as shown in Figure 7.9. The separation of adjacent interference maxima is equal to $\lambda/2$ where λ is the wavelength and hence the value of λ may be determined.

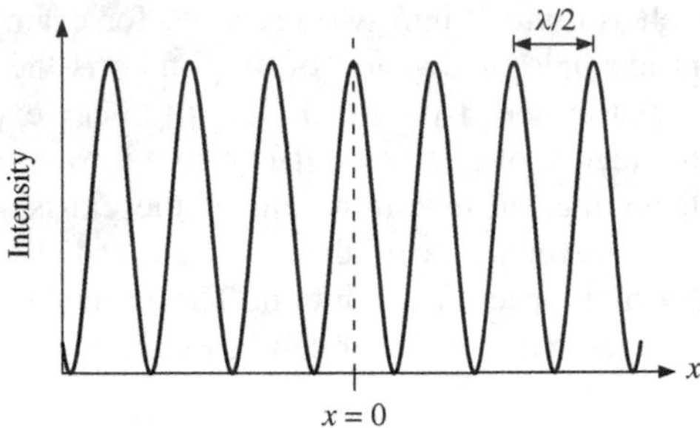

Figure 7.9 The interference pattern observed with a Michelson spectral interferometer. The measured light intensity is plotted as a function of the displacement x of the moveable mirror M_2. The separation of the maxima in the measured intensity is equal to $\lambda/2$, where λ is the wavelength of the light.

7.2 DIFFRACTION

A wave spreads out from its source becoming a plane wave at large distances, as we saw in Section 5.8. Any obstacle in the path of the wave affects the way it spreads out; the wave appears to 'bend' around the obstacle. Similarly, the wave spreads out beyond any aperture that it meets. Such bending or spreading of the wave is called diffraction. The effects of diffraction are evident in the shadow of an object that is illuminated by a point source. The edges of the shadow are not sharp but are blurred due to the bending of the light at the edges of the object. For the same reason the letters on a car number plate become blurred when we view the car from a distance of more than a few hundred metres or so. The light striking our eye bends at the iris so that the image on the retina becomes blurred. On a larger scale, waves from the Atlantic Ocean spread out after passing through the gap between Gibraltar and Spain. This is visible on satellite images of the Strait of Gibraltar, an example of which is shown in Figure 7.10. (This image was taken by a satellite of the European Space Agency.)

We shall see that the degree of spreading of a wave after passing through an aperture depends on the ratio of the wavelength λ of the wave to the size d of the aperture. The angular width of the spreading is approximately equal to λ/d; the bigger this ratio, the greater is the spreading. We begin by discussing diffraction at a single slit. This is the archetypal example of diffraction and displays all the essential physical principles.

7.2.1 Diffraction at a single slit

In our discussion of Young's double-slit experiment, we considered the width of each slit to be very narrow. This allowed us to assume that the path lengths from all points across a slit to a distant point P were equal. In practice a real slit is not arbitrarily narrow but has a finite extent. Hence, the path lengths from different points across the slit to the point P will be different and consequently the secondary wavelets arriving at P will have a variation in phase. This variation in phase gives rise to the diffraction pattern of the slit.

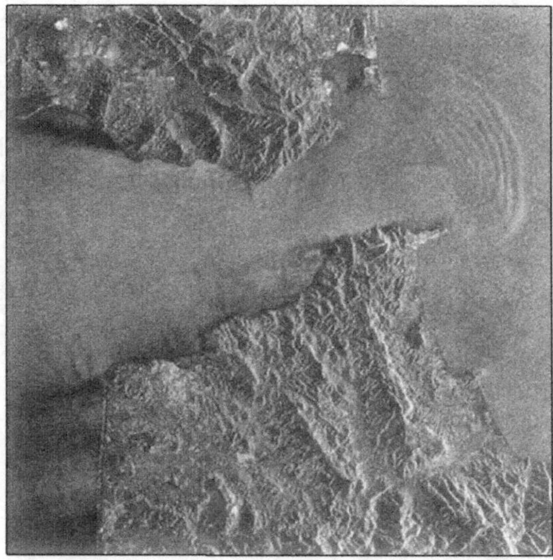

Figure 7.10 A satellite image of the Strait of Gibraltar showing the spreading of Atlantic Ocean waves after passing through the gap between Spain and Gibraltar. Image courtesy of the European Space Agency.

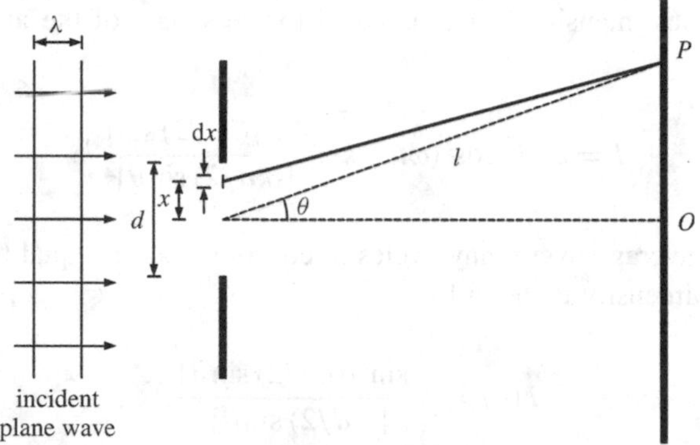

Figure 7.11 Diffraction at a single slit. The vertical scale has been enlarged for the sake of clarity. A monochromatic plane wave of wavelength λ is incident upon an opaque barrier containing a single slit. The slit has a width d and a long length ($\gg d$) in the direction normal to the page, reducing this to a two-dimensional problem. The resultant amplitude at point P is due to the superposition of secondary wavelets from the slit.

Figure 7.11 shows a monochromatic plane wave of wavelength λ that is incident on a single slit in an opaque barrier. The slit has width d and a long length ($\gg d$) in the direction normal to the page, making this a two-dimensional problem. The centre of the slit is at $x = 0$. We divide the slit into infinitely narrow strips of width $\mathrm{d}x$. Following Huygen's principle, each of these strips acts as a source of secondary wavelets and the superposition of these wavelets gives the resultant amplitude at point P. We consider the case in which P is very distant from the slit. Consequently, all the wavelets arriving at P can be assumed to be plane waves and to have the same amplitude. In addition, we can assume that the lines joining P to all points on the slit make the same angle θ to the horizontal axis.

The amplitude $\mathrm{d}R$ of the wavelet arriving at P from the strip $\mathrm{d}x$ at x is proportional to the width $\mathrm{d}x$ of the strip, and its phase depends on the distance of P from the strip, i.e. on $(l - x\sin\theta)$, where l is the distance of P from the midpoint of the slit. Hence $\mathrm{d}R$ is given by

$$\mathrm{d}R = \alpha \mathrm{d}x \cos[\omega t - k(l - x\sin\theta)], \tag{7.21}$$

where ω and k are the angular frequency and wavenumber, respectively, and α is a constant. The resultant amplitude at P due to the contributions of the secondary wavelets from all the strips is

$$R = \int_{-d/2}^{d/2} \alpha \mathrm{d}x \cos[\omega t - k(l - x\sin\theta)]. \tag{7.22}$$

We can evaluate this integral to obtain

$$R = \frac{\alpha d}{(kd/2)\sin\theta} \sin[(kd/2)\sin\theta]\cos(\omega t - kl). \tag{7.23}$$

The instantaneous intensity I at P is equal to the square of the amplitude R and thus

$$I = \alpha^2 d^2 \cos^2(\omega t - kl)\frac{\sin^2[(kd/2)\sin\theta]}{[(kd/2)\sin\theta]^2}. \tag{7.24}$$

Since the time average over many cycles of $\cos^2(\omega t - kl)$ is equal to $1/2$, the time average of the intensity is given by

$$I(\theta) = I_o \frac{\sin^2[(kd/2)\sin\theta]}{[(kd/2)\sin\theta]^2}, \tag{7.25}$$

where $I_o = \alpha^2 d^2/2$ is equal to the maximum intensity of the diffraction pattern. This equation describes how an incident plane wave of wavelength λ spreads out from a single slit of width d in terms of the angle θ. The resulting diffraction pattern is shown in Figure 7.12. This figure is a plot of $I(\theta)$ against θ for a value of $kd/2 = 10\pi$. The function

$$\frac{\sin^2[(kd/2)\sin\theta]}{[(kd/2)\sin\theta]^2} = \frac{\sin^2\beta}{\beta^2} \tag{7.26}$$

with $\beta = (kd/2)\sin\theta$ is the square of a *sinc function*. It has its maximum value of unity when $\beta = 0$. The maximum intensity I_o thus occurs when $\theta = 0$. The physical interpretation of this is that the secondary wavelets from pairs of strips at positions $\pm x$, respectively, will be in phase for $\theta = 0$, resulting in maximum intensity. The

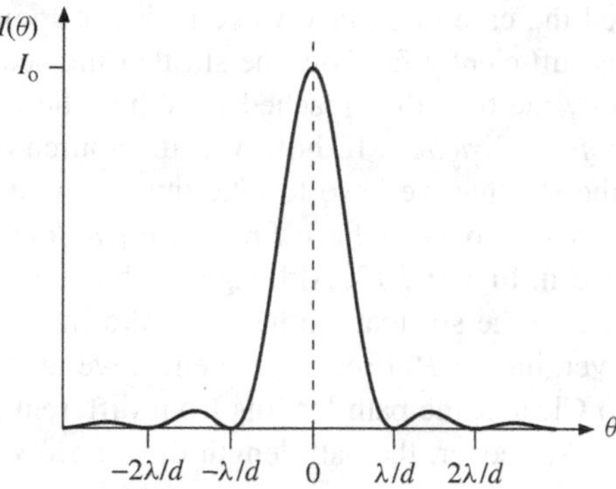

Figure 7.12 The diffraction pattern of a single slit. The intensity $I(\theta)$ is plotted against the angle θ that a line joining P to the centre of the slit makes with the horizontal as shown in Figure 7.11. The value of $\lambda/d = 0.1$. The zeros of intensity in the diffraction pattern occur at $\theta = \pm n\lambda/d$, where $n = \pm 1, \pm 2, \ldots$, under the small angle approximation $\sin\theta \simeq \theta$, which is valid in this example.

intensity $I(\theta)$ will be zero whenever the numerator of Equation (7.26), $\sin^2\beta$, is zero but the denominator, β, is not. The first zeros in the intensity occur when

$$\beta = (kd/2)\sin\theta = \pm\pi$$

and hence, using $k = 2\pi/\lambda$, when

$$\sin\theta = \pm\frac{\lambda}{d}. \tag{7.27}$$

Importantly, we see that the degree of spreading depends upon the ratio λ/d. It also depends on the wavelength which explains why we can hear sounds around a corner but we cannot see around a corner. When $\lambda \ll d$, as in the case of light, $\sin\theta$ is essentially equal to θ, giving the first zeros in the diffraction pattern at

$$\theta = \pm\frac{\lambda}{d}. \tag{7.28}$$

In general, zeros in intensity occur when

$$\theta = n\frac{\lambda}{d}, n = \pm 1, \pm 2, \ldots. \tag{7.29}$$

These zeros are shown in Figure 7.12 where the small angle approximation can be assumed since $kd/2 = 10\pi$, which gives $\lambda/d = 0.1$.

The first zeros in intensity occur for values of θ such that the path difference between the two ends of the slit is equal to one complete wavelength. We can understand this in the following way. Imagine the single slit to be composed of two slits, each of width $d/2$, placed side by side. Then the path difference between wavelets from the centres of the two slits is $\lambda/2$, which is the condition for destructive interference. Similarly, other corresponding pairs of points on the two slits will lead to destructive interference.

We have considered the case of a plane wave incident upon a single slit. Moreover, the point P was sufficiently far from the slit that the secondary wavelets had become plane waves by the time they reached P. When these conditions are satisfied we have *Fraunhofer diffraction*. If, however, the source of the primary waves or P is so close to the slit that we have to take into account the curvature of the incoming *or* outgoing wavefronts we have *Fresnel diffraction*. The case of Fresnel diffraction is illustrated in Figure 7.13. Although we have an incident plane wave, the point P is so close to the slit that we have to take into account the curvature of the wavefront converging on P. (For convenience, we take P to be in line with the centre of the slit.) Clearly, the path lengths from different points across the slit to P will be different. Moreover, the path-length difference s at a distance x from

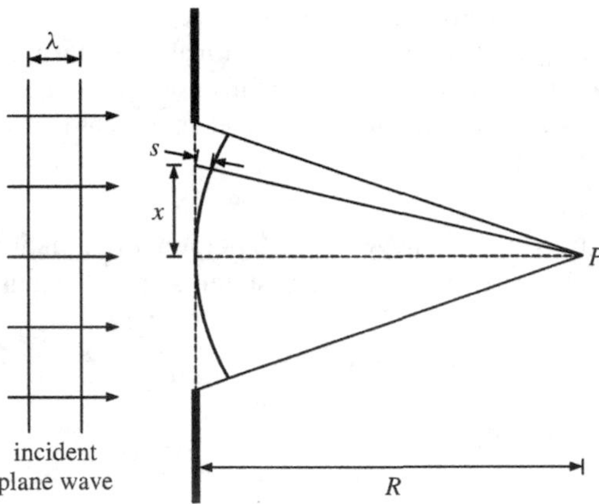

Figure 7.13 A plane wave is incident on a single slit. Point P is so close to the slit that the curvature of the wave converging on P has to be taken into account in determining the resultant amplitude at that point. This is an example of Fresnel diffraction.

the centre of the slit is not linearly proportional to x. It is easy to show that s is given by

$$s \simeq \frac{x^2}{2R}, \tag{7.30}$$

when $x^2/R^2 \ll 1$. Hence the phase difference $\phi(x)$ for a point at x is

$$\phi(x) \simeq \frac{2\pi}{\lambda}\frac{x^2}{2R}, \tag{7.31}$$

where λ is the wavelength. The phase difference has a *quadratic* dependence on position x, which is a characteristic of Fresnel diffraction. This is in contrast to Fraunhofer diffraction where we found that the path difference is linearly proportional to x. [The path difference is equal to $x \sin\theta$, where θ is the direction of the secondary wavelets, cf. Equation (7.21).] There is no sharp division between Fraunhofer and Fresnel diffraction, the pattern changes continuously from one to the other as the distance from the slit to P reduces. To illustrate the transition

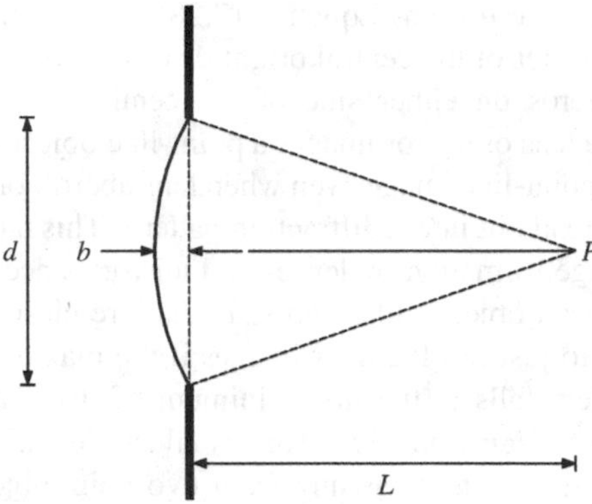

Figure 7.14 In Fraunhofer diffraction we require the curvature of the wavefront at a slit to be sufficiently large that the wavefront can be considered to be planar, i.e. that the distance L is sufficiently large that $b \ll \lambda$.

from Fraunhofer to Fresnel diffraction, Figure 7.14 shows a circular wavefront that converges on a point P that is at a distance L from the slit. It is easy to show that

$$b \simeq \frac{d^2}{8L} \tag{7.32}$$

for $d^2/L^2 \ll 1$, where b is the distance between the circular wavefront and the plane of the slit as shown. For Fraunhofer diffraction we require the curvature of the wavefront to be sufficiently large that the wavefront can be considered to be planar, i.e. that b be much less than the wavelength λ. Hence, we must have

$$L \gg \frac{d^2}{8\lambda}. \tag{7.33}$$

When L reduces, so that it becomes comparable with $d^2/8\lambda$, we have Fresnel diffraction.

7.2.2 Circular apertures and angular resolving power

A circular aperture will also produce a diffraction pattern. This pattern will, of course, have circular symmetry. For a plane wave that falls normally on a small circular aperture, the diffraction pattern appears as a central bright disc surrounded by a series of bright and dark rings. This central disc is called the *Airy disc* in honour of Sir George Airy, a former Astronomer Royal of England, and contains 84% of the integrated light intensity. The dark rings correspond to the zeros of intensity in the diffraction pattern. For an aperture of diameter d, the first zeros on either side of the central maximum occur at angles $\pm\theta_R$, where

$$\theta_R = 1.22\frac{\lambda}{d}. \tag{7.34}$$

This equation has the same form as Equation (7.28) but with the multiplying factor 1.22. The angular diameter of the central bright disc is equal to the angular distance between these two zeros on either side of the central maximum, i.e. $2.44\lambda/d$. Consequently, when a lens or mirror images a point-like object such as a distant star, it does not produce a point-like image even when lens aberrations can be discounted. Instead the light is spread out into a diffraction pattern. This has important practical consequences for image formation by lenses and mirrors since it limits their ability to resolve closely spaced objects, like two stars that are close together in our field of view. We are able to just resolve their images if the maximum of the diffraction pattern from one object falls at the first minimum of the pattern from the other, accordingly to the *Rayleigh criterion*. This is illustrated in Figure 7.15, which shows the two diffraction patterns arising from two point objects. The dotted line is the sum of the two diffraction patterns and illustrates that we can just distinguish the two diffraction maxima. It follows that we would just be able to distinguish the two point images. The angular separation of two *objects* is the same as the angular separation of their *images*. Hence, two point objects are just resolvable by a lens or mirror of diameter d when their angular separation θ satisfies

$$\boxed{\theta = 1.22\frac{\lambda}{d}.} \tag{7.35}$$

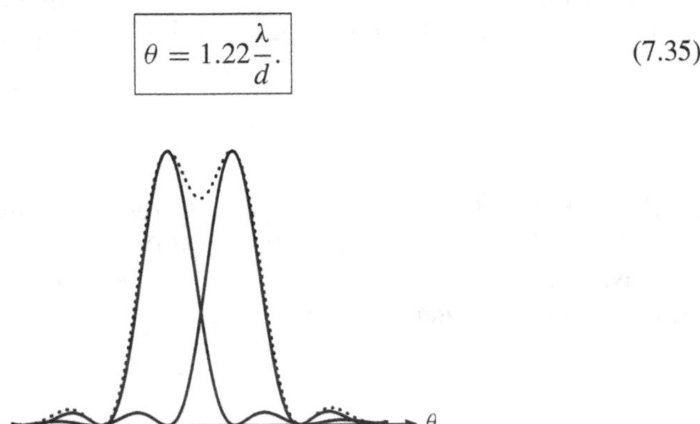

Figure 7.15 Two overlapping diffraction patterns at the image plane of a lens or mirror, arising from two point objects. The dotted line is the sum of the two diffraction patterns. The *Rayleigh criterion* states that the images of the two point objects can be just resolved when the maximum of one diffraction pattern overlaps the first minimum of the other. This is the case shown.

If two objects with a spatial separation b are at a large distance L from a lens or mirror, then we can write $\theta = b/L$. Hence we can just resolve them if $b = 1.22\lambda L/d$. For example, if we assume a size of 3 mm for a human pupil and an optical wavelength of 550 nm, we have

$$\theta = \frac{b}{L} = 1.22\frac{\lambda}{d} = 1.22\frac{550 \times 10^{-9}}{3 \times 10^{-3}} \approx \frac{1}{5000}.$$

This suggests that we can read a car number plate at a distance of $\sim$100 m, assuming that we need to resolve features $\sim$2 cm apart. In radio astronomy the wavelengths of interest are much longer than for visible light. For example, atomic hydrogen produces what is known as *21 centimetre radiation* and this is used extensively in radio astronomy. The diameter of the Lovell telescope at Jodrell Bank, UK, is 76 m. At the wavelength of 21 cm, it has an angular resolution ~1/300 for which $\theta \sim 0.2°$. Similar considerations apply to microscopy. In an *electron microscope* the wavelengths associated with the electrons may be 100,000 times shorter than for visible light and so sharp images of extremely small objects can be obtained. Diffraction also limits the amount of information that can be stored on optical recording media like compact discs. There is no point in making the dimensions of the pattern printed on the disc smaller than the diffraction limit of the optical imaging system that is used to read it.

7.2.3 Double slits of finite width

We are now in a position to take into account the finite width of the slits in a real Young's double-slit experiment. As for the analysis of diffraction at a single slit, we consider each of the two slits to be composed of infinitely narrow strips that act as sources of secondary wavelets. Then the resultant amplitude R at a point P is the superposition of the secondary wavelets from both slits. This is given by

$$R = \int_{-a/2-d/2}^{-a/2+d/2} \alpha \mathrm{d}x \cos[\omega t - k(l - x\sin\theta)] + \int_{a/2-d/2}^{a/2+d/2} \alpha \mathrm{d}x \cos[\omega t - k(l - x\sin\theta)], \tag{7.36}$$

where d is the width of each slit and a is their separation, cf. Equation (7.22). Evaluating these integrals gives

$$R = 2\alpha d \cos(\omega t - kl)\frac{\sin[(kd/2)\sin\theta]}{(kd/2)\sin\theta}\cos[(ka/2)\sin\theta]. \tag{7.37}$$

The resultant intensity is

$$I(\theta) = I_{\text{o}}\frac{\sin^2[(kd/2)\sin\theta]}{[(kd/2)\sin\theta]^2}\cos^2[(ka/2)\sin\theta], \tag{7.38}$$

where I_{o} is the maximum intensity of the pattern. This result is the product of two functions. The first is the square of a sinc function corresponding to diffraction at a single slit, cf. Equation (7.25). The second is the cosine-squared term of the double-slit interference pattern, cf. Equation (7.10). These two functions are displayed separately in Figure 7.16(b) and (a), respectively. The physical interpretation of Equation (7.38) is that the double-slit interference pattern is modulated by the intensity pattern due to diffraction of the incoming plane wave at each slit. The result of this modulation is shown in Figure 7.16(c), which is the interference pattern for two slits of finite width. Both of the above functions, i.e. the cosine squared and sinc squared functions, have maxima and minima at particular values of θ. In particular, and for the small angle approximation

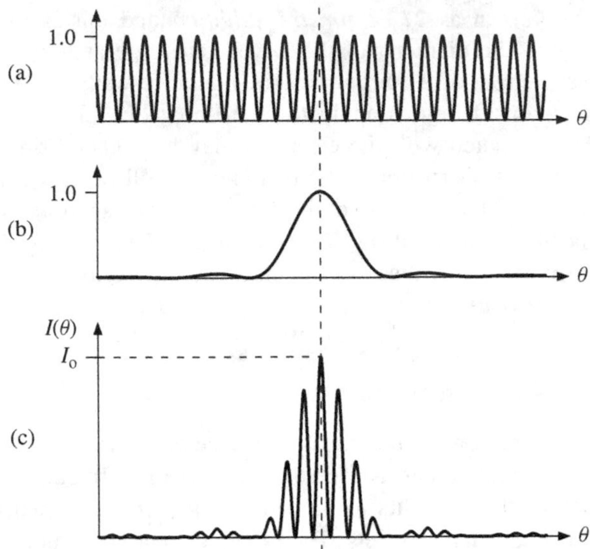

Figure 7.16 (a) The cosine squared term appearing in Equation (7.38) corresponding to interference fringes in a Young's double-slit experiment with infinitely narrow slits, cf. Equation (7.10). (b) The sinc squared function appearing in Equation (7.38) corresponding to diffraction at a single slit, cf. Equation (7.25). (c) The observed interference pattern from a Young's double-slit experiment with slits of finite width; corresponding to the modulation of the cosine squared term in (a) by the sinc squared function in (b). The small angle approximation, $\sin\theta \simeq \theta$, has been used.

$\sin\theta \simeq \theta$ used in Figure 7.16, double-slit interference maxima occur at angles given by

$$\theta = \frac{n\lambda}{a}, \quad n = 0, \pm 1, \pm 2, \ldots, \qquad \text{cf. (7.12)}$$

while zeros in the diffraction pattern occur at angles given by

$$\theta = \frac{n\lambda}{d}, n = 0, \pm 1, \pm 2, \ldots. \qquad \text{cf. (7.29)}$$

Clearly if an interference maximum occurs at a zero in the diffraction pattern, that bright fringe will be absent from the observed pattern. In the example shown in Figure 7.16 the ratio $a : d = 4 : 1$ and consequently the $n = 4$ bright fringe is missing.

We obtained Equation (7.38) by considering the diffraction pattern observed for two slits of finite width. However, it is an example of a more general result: the diffraction pattern from a system consisting of any number of slits will always have an envelope corresponding to single slit-diffraction modulating the multi-slit interference pattern. This occurs, for example, in the case for a diffraction grating.

PROBLEMS 7

(Take the velocity of sound in air to be 340 m s^{-1}.)

7.1 (a) In a Young's double-slit experiment, it is found that ten bright interference fringes span a distance of 1.8 cm on a screen placed 1.0 m away. The separation of the two slits is 0.30 mm. Determine the wavelength of the light. (b) Light from a helium-neon laser with wavelength 633 nm is incident upon two very narrow slits spaced 0.50 mm apart. The viewing screen is placed a distance of 1.5 m beyond the slits. What are the distances between (i) the two $n = 2$ bright fringes and (ii) the two $n = 2$ dark fringes?

7.2 In a Young's double-slit experiment, the angular separation of the interference fringes on a distance screen is 0.04°. What would be the angular separation if the entire apparatus were immersed in a liquid of refractive index 1.33?

7.3 Plane waves of monochromatic light of wavelength 500 nm are incident upon a pair of very narrow slits producing an interference pattern on a screen. When one of the slits is covered by a thin film of transparent material of refractive index 1.60 the central ($n = 0$) bright fringe moves to the position previously occupied by the $n = 15$ bright fringe. What is the thickness of the film?

7.4 (a) Estimate the divergence angle of the sunlight we receive on Earth given that the diameter of the Sun is 1.4×10^6 km and its distance from the Earth is 1.5×10^8 km. (b) In a Young's double-slit experiment, the slit spacing is 0.75 mm and the wavelength of the incident light is 550 nm. What should be the maximum divergence angle of the source for the interference fringes to be clearly visible? Compare this value with your answer from (a).

7.5 The two slits in a Young's double-slit experiment each have a width of 0.06 mm and are separated by a distance a. If an $n = 15$ bright fringe of the double-slit interference pattern falls at the first minimum of the diffraction pattern due to each slit, what is the value of the separation of the slits a?

7.6 Two loudspeakers are separated by a distance of 1.36 m. They are connected to the same amplifier and emit sound waves of frequency 1.0 kHz. How many maxima in sound intensity would you hear if you walked in a complete circle around the loudspeakers at a large distance from them? Assume that the sound waves are emitted isotropically.

7.7 (a) Monochromatic light is directed into a Michelson spectral interferometer. It is observed that 4001 maxima in the detected light intensity span exactly 1.0 mm of mirror movement. What is the wavelength of the light? (b) Light from a sodium discharge lamp is directed into a Michelson spectral interferometer. The light contains two wavelength components having wavelengths of 589.0 nm and 589.6 nm, respectively. The interferometer is initially set up with its two arms of equal length so that a maximum in the detected light is observed. How far must the moveable mirror be moved so that the 589.0 nm component produces one more maximum in the detected intensity than the 589.6 nm component?

7.8 A gas cell of length 8.0 cm is inserted into the light path in one of the arms of a Michelson spectral inteferometer. Light from a helium-neon laser with wavelength 633 nm is directed into the interferometer. Initially the gas cell is evacuated of air and the interferometer is adjusted for maximum intensity at the detector. Air is then slowly leaked into the gas cell until the pressure reaches atmospheric pressure. As this is done it is found that the light intensity at the detector passes from maximum to minimum intensity and back to maximum intensity exactly 90 times. Use these data to determine the refractive index of air at atmospheric pressure.

7.9 If you clap your hands at the centre of a Roman amphitheatre, you may hear a sound similar to that produced by a plucked string. Explain this phenomenon and estimate the frequencies involved.

7.10 (a) A car is travelling towards you on a long straight road at night. Estimate the distance at which you can just resolve its headlights into two separate sources of light. Would the light from the two separate headlights produce any interference effects? (b) The Hubble Space Telescope has a diameter of 2.4 m. Determine its diffraction-limited angular resolution at a wavelength of 550 nm in radians and in degrees.

7.11

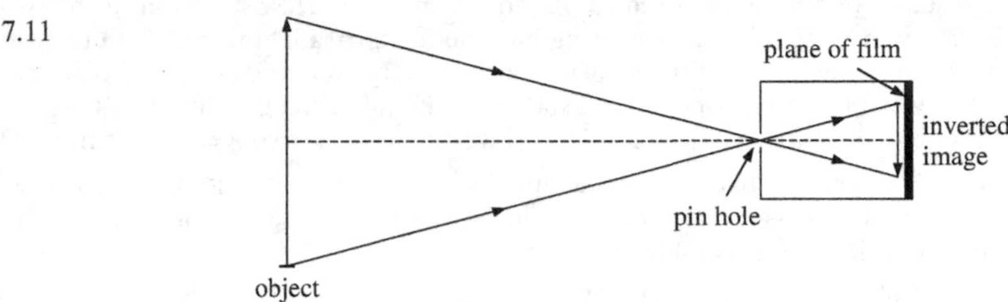

The figure illustrates the principle of operation of a pin-hole camera, which produces an inverted image of an object at the plane of the film. This image becomes blurred if the diameter d of the pin hole is too large or too small. (a) Explain why this blurring occurs for both the above cases. (b) The pin hole has an optimum diameter when the two effects above give rise to the same amount of blurring. Show that for distant objects, the optimum value of $d \approx \sqrt{2.44\lambda l}$, where λ is the wavelength and l is the distance between the pin hole and the plane of the film. (c) Using an appropriate value of λ, evaluate the optimum value of d for $l = 15$ cm.

8

The Dispersion of Waves

In our discussion of waves so far, we have considered the velocity of a wave to be independent of its frequency. In some important cases this is true. The velocity of electromagnetic waves in a vacuum is independent of frequency. To a very good approximation, the velocity of sound waves in air is also independent of frequency. This is just as well since otherwise the members of the audience sitting at the back of an auditorium would have a very different musical experience to those sitting at the front. And, in our discussion of transverse waves on a taut string, we found that the velocity of the waves, $v = \sqrt{T/\mu}$ is independent of frequency. In general, however, the velocity of a wave in a medium does depend on its frequency. This is called *dispersion* and the medium in which the wave travels is called a *dispersive medium*. A familiar example of this is the separation of white light into the colours of the rainbow by a glass prism. The light is dispersed because the velocity of light in glass varies with frequency. In many situations, we do not deal with a single wave but rather with a *group* of waves having different frequencies. The superposition of these waves leads to a *modulated* wave. In a dispersive medium, the individual waves in the group travel at different velocities and change their relative positions as they propagate. Consequently, the modulation of the wave travels at a velocity, called the *group velocity*, which is different from the velocities of the waves in the group. We first consider, in Section 8.1, the superposition of waves and their propagation in non-dispersive media. In Section 8.2 we extend our discussion to the propagation of waves in dispersive media.

8.1 THE SUPERPOSITION OF WAVES IN NON-DISPERSIVE MEDIA

The travelling wave $\psi = A\cos(kx - \omega t)$ is described as monochromatic because it has a single frequency ω and a single wavelength λ $(= 2\pi/k)$. Moreover, it extends to $\pm$ infinity along the x-axis. (In practice this is unrealistic and a real wave has a beginning and an end, although its length may be considerable. For example,

Vibrations and Waves George C. King

researchers have used lasers to produce monochromatic light waves many kilometres long.) A monochromatic wave cannot carry any information since its amplitude and frequency do not vary. To send information we need to *modulate* the wave in some way as is done, for example, in the transmission of Morse code. A modulated wave consists of the superposition of a group of waves of different frequencies. We have already met the superposition of waves in, for example, the formation of standing waves. There the waves travelled in opposite directions. Here we consider the superposition of waves travelling in the same direction. We shall consider the phenomenon of *beats* and also the *amplitude modulation* of radio waves where it is clearly the intention to transmit information.

8.1.1 Beats

The simplest superposition we can have consists of two monochromatic waves

$$\psi_1 = A\cos(k_1 x - \omega_1 t), \qquad \psi_2 = A\cos(k_2 x - \omega_2 t), \tag{8.1}$$

that have the same amplitude A but different frequencies ω_1 and ω_2, respectively. In a non-dispersive medium, the two waves travel at the same velocity:

$$v = \frac{\omega_1}{k_1} = \frac{\omega_2}{k_2}. \tag{8.2}$$

The superposition of the two waves gives

$$\psi = \psi_1 + \psi_2 = A\cos(k_1 x - \omega_1 t) + A\cos(k_2 x - \omega_2 t). \tag{8.3}$$

Using the identity

$$\cos(\alpha + \beta) + \cos(\alpha - \beta) = 2\cos\alpha\cos\beta \tag{8.4}$$

and letting

$$(\alpha + \beta) = (k_2 x - \omega_2 t), \qquad (\alpha - \beta) = (k_1 x - \omega_1 t) \tag{8.5}$$

we obtain

$$\psi = 2A\cos\left[\frac{(k_2 - k_1)}{2}x - \frac{(\omega_2 - \omega_1)}{2}t\right]\cos\left[\frac{(k_2 + k_1)}{2}x - \frac{(\omega_2 + \omega_1)}{2}t\right]. \tag{8.6}$$

We consider how ψ varies at a fixed value of position x. This would be the situation, for example, where a superposition of two sound waves impinges on our eardrum. For convenience we take $x = 0$, so that Equation (8.6) becomes

$$\psi = 2A\cos\left[\frac{(\omega_2 - \omega_1)}{2}t\right]\cos\left[\frac{(\omega_2 + \omega_1)}{2}t\right]. \tag{8.7}$$

The result is the product of two cosine terms with frequencies of $(\omega_2 - \omega_1)/2$ and $(\omega_2 + \omega_1)/2$, respectively. This is a general result that applies to any two frequencies ω_2 and ω_1. However, this result is particularly interesting when the two frequencies are nearly the same, i.e. $\omega_1 \approx \omega_2$. We then have a wave of frequency $(\omega_2 + \omega_1)/2$ that is multiplied, i.e. modulated, by a term that varies much more slowly, since $(\omega_2 - \omega_1)/2 \ll (\omega_2 + \omega_1)/2$. This situation is illustrated by Figure 8.1(a) which shows the two monochromatic waves, and Figure 8.1(b) which shows their superposition. We see that the waves sometimes add constructively and sometimes destructively because of their different frequencies. This phenomenon is called *beats*. The resultant wave is contained within an *envelope* shown by the dotted lines in Figure 8.1(b). The envelope is periodic as given by Equation (8.7) with the two dotted lines being defined by $\pm 2A\cos[(\omega_2 - \omega_1)t/2]$. We can rewrite Equation (8.7) in the form

$$\psi = A(t)\cos\omega_{\rm o}t, \tag{8.8}$$

where $\omega_{\rm o} = (\omega_2 + \omega_1)/2$ and the amplitude $A(t)$ is given by

$$A(t) = 2A\cos\left[\frac{(\omega_2 - \omega_1)}{2}t\right]. \tag{8.9}$$

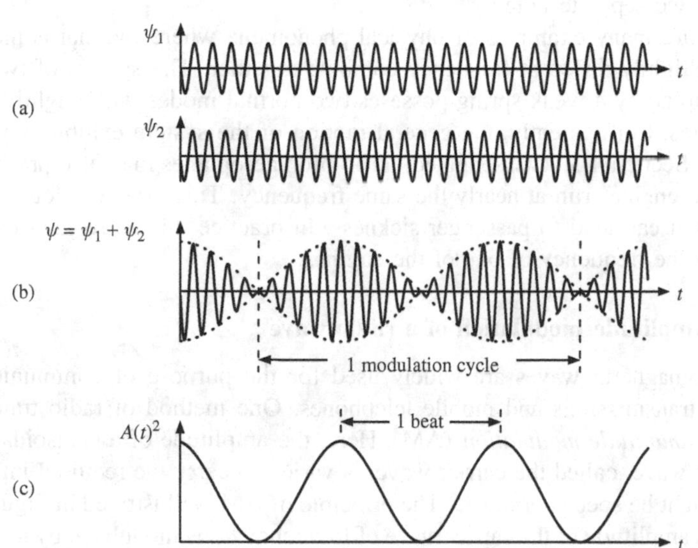

Figure 8.1 (a) Two monochromatic waves ψ_1 and ψ_2, having the same amplitude but slightly different frequencies. (b) The superposition ψ of the two waves showing the resulting beat pattern. (c) The square of the amplitude of the modulation $A(t)^2$, which reaches a maximum value twice during each period of the beat pattern.

The wave described by Equation (8.8) is not a true sinusoidal wave since its amplitude varies. However, under the condition that $\omega_1 \approx \omega_2$, the variation will be slow and there will be many high frequency oscillations within each period of the envelope, as in the example of Figure 8.1(b). It is reasonable then to describe Equation (8.8) as a sinusoidal wave of frequency $\omega_{\rm o}$, although one with a slowly varying amplitude.

An example of beats occurs when we simultaneously strike two tuning forks that have slightly different frequencies. We hear a note with a well defined pitch but with a sound intensity that rises and falls periodically. In this example ψ_1 and ψ_2 represent the two sound waves emitted by the tuning forks where each is a measure of the pressure variation in the air. ψ is the superposition of the two sound waves. The intensity, or loudness, of the sound is proportional to ψ^2 and hence is proportional to $A(t)^2$, which is shown in Figure 8.1(c). The frequency of the modulation is $(\omega_2 - \omega_1)/2$, Equation (8.7). However, $A(t)^2$ reaches a maximum twice during each period of the modulating term. It follows that the sound will reach maximum intensity at twice the frequency of the modulation and so the beat frequency is just the difference between the frequencies of the two tuning forks. For example, if we had one fork tuned to 439 Hz and the other to 401 Hz, we would hear a note of frequency 440 Hz and a beat frequency of 2 Hz. The method of beats is commonly used to tune string instruments. A string of the instrument is plucked while a tuning fork of the required frequency is struck simultaneously. Beats are heard if the two are slightly out of tune. Tuning is accomplished by adjusting the tension in the string until the beat frequency reduces to zero. A person can discern beats up to a maximum frequency of about 5–10 Hz. Above this, the sound is heard as two separate notes.

There are many examples of physical phenomena where two harmonic oscillations of slightly different frequency combine together. The system of two pendulums coupled by a weak spring posseses two normal modes with slightly different frequencies. Consequently, the general motion of the system exhibits a pattern of beats, cf. Section 4.3. Twin-engined, turbo-prop aeroplanes may also produce beats if the two engines run at nearly the same frequency. This produces loud throbbing sounds that can lead to passenger sickness. In practice this is avoided by slightly changing the frequency of one of the engines.

8.1.2 Amplitude modulation of a radio wave

Electromagnetic waves are widely used for the purpose of communication as in radio transmissions and mobile telephones. One method of radio transmission employs *amplitude modulation* (AM). Here, the amplitude of a sinusoidal electromagnetic wave, called the carrier wave, is varied to carry the required information which might be speech or music. The principle of AM is illustrated in Figure 8.2(a). Here, the amplitude of the carrier wave of frequency ω_c is modulated by a sinusoidal wave of much lower frequency ω_m. The resultant wave can be represented by

$$\psi = (A + B\cos\omega_m t)\sin\omega_c t. \tag{8.10}$$

B is called the *depth of modulation*, which must be less than A to avoid distortion of the signal at the receiver. Using the trigonometric identity

$$\sin\alpha\cos\beta = \frac{1}{2}[\sin(\alpha+\beta) + \sin(\alpha-\beta)] \tag{8.11}$$

we can rewrite Equation (8.10) as

$$\psi = A\sin\omega_c t + \frac{B}{2}[\sin(\omega_c + \omega_m)t + \sin(\omega_c - \omega_m)t]. \tag{8.12}$$

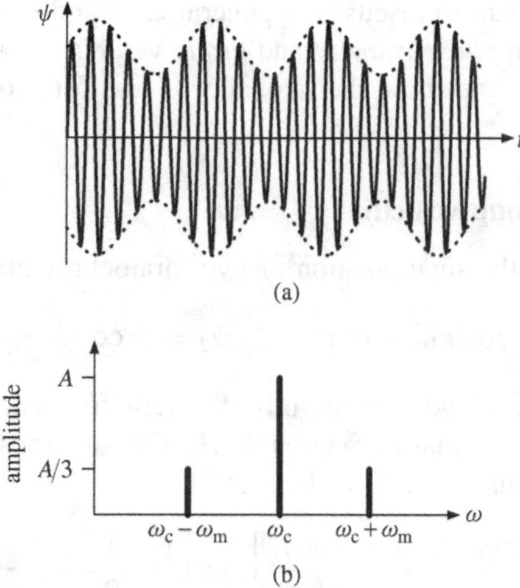

Figure 8.2 (a) The principle of AM radio transmission. A carrier wave of frequency ω_c is modulated by a sinusoidal wave of frequency ω_m, where $\omega_m \ll \omega_c$. The resultant waveform ψ is shown for $x = 0$. (b) The frequency spectrum of the modulated carrier wave showing the frequency components present.

Inspection of Equation (8.12) shows that there are three frequency components present in the modulated wave. These are the carrier frequency ω_c and the frequencies $(\omega_c + \omega_m)$ and $(\omega_c - \omega_m)$. We can represent these components as a frequency spectrum as shown in Figure 8.2(b). In this spectrum the heights of the lines represent the amplitudes of the frequency components and in this particular example, $B = A/3$. Of course a real audio signal contains a continuous range of frequencies, typically 10 Hz to 10 kHz, and so ω_m will have this range also. (This compares with the carrier frequency which is typically ~1 MHz, i.e. $\omega_c \gg \omega_m$.) Consequently there is a band of frequencies on either side of the central frequency ω_c, which are called *side bands*. It follows that adjacent radio stations must have carrier frequencies that differ by more than $2\omega_m$. (In more sophisticated AM transmission systems, only the frequencies of a single side band are transmitted so that more radio stations can fit into the available frequency range.)

8.2 THE DISPERSION OF WAVES

In a non-dispersive medium, the velocity of a wave is independent of the wavenumber k, i.e. $v = \omega/k =$ constant, and

$$\omega = \text{constant} \times k.$$

In a dispersive medium the velocity $v = \omega/k$ does depend on the wavenumber k, and so also will the frequency $\omega = vk$. The relationship between the frequency ω and the wavenumber k is called the *dispersion relation* of the medium. The dispersion relation is determined by the physical properties of the medium. Different media will, in general, have different dispersion relations and these will lead to different wave behaviours. In Section 8.3 we shall illustrate these different types of behaviour. Here we want to discuss the general case and in particular to illustrate the difference between *phase velocity* and *group velocity*. For this it will suffice to note that in a dispersive medium the frequency ω is a function of the wavenumber k: $\omega = \omega(k)$.

8.2.1 Phase and group velocities

We again consider the superposition of two monochromatic waves:

$$\psi_1 = A\cos(k_1 x - \omega_1 t), \qquad \psi_2 = A\cos(k_2 x - \omega_2 t), \tag{8.1}$$

that have the same amplitude but slightly different frequencies, so that $\omega_1 \approx \omega_2$. The analysis is similar to that of Section 8.1.1. The superposition of ψ_1 and ψ_2 is given by the same Equation (8.6) as before:

$$\psi = 2A\cos\left[\frac{(k_2 - k_1)}{2}x - \frac{(\omega_2 - \omega_1)}{2}t\right]\cos\left[\frac{(k_2 + k_1)}{2}x - \frac{(\omega_2 + \omega_1)}{2}t\right]. \tag{8.6}$$

The difference here is that the medium is dispersive and so the two waves have different velocities given by $v_1 = \omega_1/k_1$ and $v_2 = \omega_2/k_2$, respectively. We let

$$k_o = \frac{(k_2 + k_1)}{2}, \qquad \omega_o = \frac{(\omega_2 + \omega_1)}{2} \tag{8.13}$$

where k_o and ω_o are the mean values of the wave numbers and frequencies, respectively. Since the differences between ω_1 and ω_2 and between k_1 and k_2 are small, we write

$$\frac{(k_2 - k_1)}{2} = \Delta k, \qquad \frac{(\omega_2 - \omega_1)}{2} = \Delta\omega. \tag{8.14}$$

In this case, Equation (8.6) can be written as

$$\psi = A(x, t)\cos(k_o x - \omega_o t) \tag{8.15a}$$

where

$$A(x, t) = 2A\cos(x\Delta k - t\Delta\omega). \tag{8.15b}$$

Equation (8.15a) represents a wave that has a frequency ω_o, a wavenumber k_o and velocity v given by

$$v = \frac{\omega_o}{k_o}. \tag{8.16}$$

The velocity v is called the wave or *phase* velocity. The amplitude of the wave $A(x, t)$ is modulated according to Equation (8.15b) and this modulation forms an envelope that contains the wave. This envelope is represented by the dotted lines in Figure (8.3). The envelope also travels forward with the wave but it does so with a velocity that, in general, is different from the phase velocity of the wave. A crest of the *envelope* will travel at the envelope velocity, as depicted by the bold dots in Figure (8.3). The amplitude of this crest remains constant as the envelope travels along, i.e. the crest maintains a constant value of modulation amplitude $A(x, t)$.

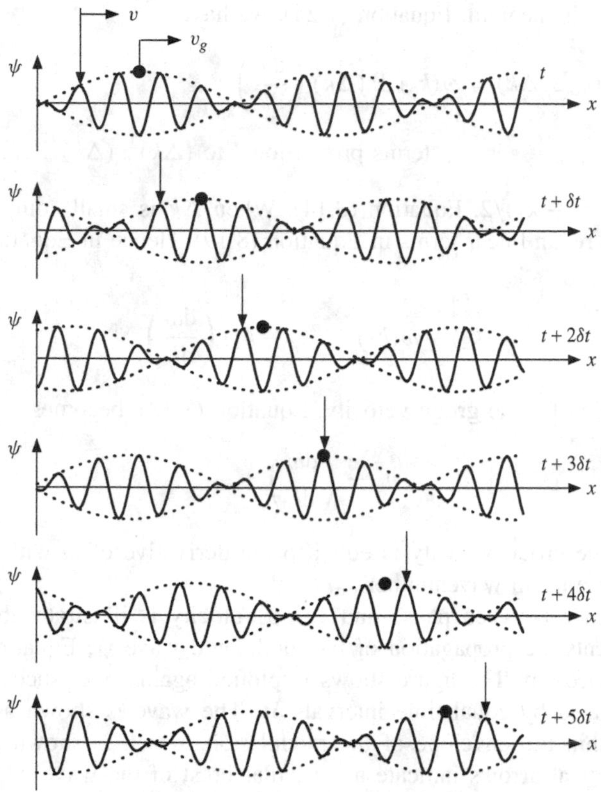

Figure 8.3 The propagation of the modulated wave ψ in a dispersive medium. ψ is plotted against x at successive, equal intervals of time δt. The wave is shown as a solid line and is contained within the envelope of the modulation, which is represented by the dotted lines. The vertical arrows indicate a particular crest of the wave that travels at the phase velocity v. The bold black dots indicate a particular crest of the envelope that travels at the group velocity v_g. In this example $v > v_g$ and so the wave crest moves forward through the envelope as the wave propagates, as can be seen from the changing relative positions of the bold dots and arrows.

From Equation (8.15b), the condition $A(x, t) = \text{constant}$, reduces to

$$x\Delta k - t\Delta\omega = \text{constant}$$

Differentiating this equation with respect to t, we obtain the velocity at which the envelope travels:

$$v_g \equiv \frac{\mathrm{d}x}{\mathrm{d}t} \simeq \frac{\Delta\omega}{\Delta k} = \frac{\omega_2 - \omega_1}{k_2 - k_1}. \tag{8.17}$$

This velocity v_g is called the *group* velocity. Since ω is a function of wavenumber k in a dispersive medium, we write Equation (8.17) as

$$v_g = \frac{\omega(k_2) - \omega(k_1)}{k_2 - k_1}. \tag{8.18}$$

Using Taylor's theorem, Equation (1.24), we have

$$\omega(k_o \pm \Delta k) = \omega(k_o) \pm (\Delta k)\left(\frac{d\omega}{dk}\right)_{k=k_o} + \text{terms proportional to } (\Delta k)^2, (\Delta k)^3, \ldots, \tag{8.19}$$

where $\Delta k = (k_2 - k_1)/2$, Equation (8.14). When Δk is small compared with k_o, we need only retain linear terms in Equation (8.19). Hence in Equation (8.18), we can write

$$\omega(k_2) - \omega(k_1) = (k_2 - k_1)\left(\frac{d\omega}{dk}\right)_{k=k_o} \tag{8.20}$$

and the equation for the group velocity, Equation (8.18), becomes

$$v_g = \left(\frac{d\omega}{dk}\right)_{k=k_o}. \tag{8.21}$$

We see that the group velocity is equal to the derivative of ω with respect to k, evaluated at the mean wavenumber k_o.

The difference between phase and group velocity is illustrated by Figure 8.3 which represents the propagation of the modulated wave ψ, Equation (8.15a), in a dispersive medium. The figure shows ψ plotted against x at successive instants of time separated by equal time intervals δt. The wave is shown as a solid line contained within the envelope of the modulation, which is shown as the dotted lines. The vertical arrows indicate a particular crest of the wave which travels at the phase velocity $v = \omega_o/k_o$. The bold black dots indicate a particular crest of the envelope which travels at the group velocity, $v_g = (d\omega/dk)_{k=k_o}$. In this example $v > v_g$ and so the wave crest moves forward through the envelope as the modulated wave propagates. This can be discerned from the changing relative positions of the bold dots and arrows.[1]

We have obtained expressions for the phase and group velocities using the example of the superposition of just two monochromatic waves. These expressions, however, apply to any group of waves so long as their frequency range is narrow compared to their mean frequency. Thus for the general case, we define the phase velocity v as

$$\boxed{v = \frac{\omega}{k},} \tag{8.22}$$

and the group velocity v_g as

$$\boxed{v_g = \frac{d\omega}{dk}.} \tag{8.23}$$

A good way to observe the behaviour of a group of waves and to appreciate the difference between phase and group velocities is to make water ripples by throwing

[1] Figure 8.3 was generated using a spreadsheet program where it is straightforward to change the ratio of phase and group velocities. The reader is strongly encouraged to try this exercise.

a stone into a still pond. What we observe is a *group* of ripples expanding outwards. For such water waves, the phase velocity is greater than the group velocity. Thus each ripple appears at the rear of the envelope of the group, proceeds through it and then disappears at the front with a new ripple appearing at the rear, cf. Figure 8.3.

The expression for the group velocity, Equation (8.23), may be rewritten in various different forms. For example, since $v = \omega/k$, Equation (8.22), we have

$$v_g = \frac{\mathrm{d}\omega}{\mathrm{d}k} = \frac{\mathrm{d}(kv)}{\mathrm{d}k} = v + k\frac{\mathrm{d}v}{\mathrm{d}k} = v + k\frac{\mathrm{d}v}{\mathrm{d}\lambda}\frac{\mathrm{d}\lambda}{\mathrm{d}k}.$$

Since $k = 2\pi/\lambda$,

$$\frac{\mathrm{d}\lambda}{\mathrm{d}k} = -\frac{\lambda}{k}.$$

and hence

$$v_g = v - \lambda\frac{\mathrm{d}v}{\mathrm{d}\lambda}. \tag{8.24}$$

Usually $\mathrm{d}v/\mathrm{d}\lambda$ is positive and so $v_g < v$. This is called *normal dispersion*. *Anomalous dispersion* occurs when $\mathrm{d}v/\mathrm{d}\lambda$ is negative so that $v_g > v$. If there is no dispersion, $\mathrm{d}v/\mathrm{d}\lambda = 0$ and the group and phase velocities are equal.

Worked example

The yellow light from a sodium lamp has two components with wavelengths of 589.00 nm and 589.59 nm. The refractive index n of a particular glass at these wavelengths has the values 1.6351 and 1.6350, respectively. Determine (i) the phase velocities of the light at these two wavelengths in the glass and (ii) the velocity of a narrow pulse of sodium light that is transmitted through the glass.

Solution

(i) Since $n = c/v$:

At 589.00 nm, $v = c/1.6351 = 0.61158c$, and at 589.00 nm, $v = 0.61162c$.

(ii) The light pulse travels at the group velocity. From $n = c/v$:

$$\frac{\mathrm{d}v}{\mathrm{d}\lambda} = \frac{\mathrm{d}v}{\mathrm{d}n}\frac{\mathrm{d}n}{\mathrm{d}\lambda} = -\frac{c}{n^2}\frac{\mathrm{d}n}{\mathrm{d}\lambda} = -\frac{v}{n}\frac{\mathrm{d}n}{\mathrm{d}\lambda}.$$

Hence from Equation (8.24)

$$v_g = v - \lambda\left(-\frac{v}{n}\frac{\mathrm{d}n}{\mathrm{d}\lambda}\right) = v\left(1 + \frac{\lambda}{n}\frac{\mathrm{d}n}{\mathrm{d}\lambda}\right).$$

Taking

$$v = \frac{(0.61162 + 0.61158)c}{2},$$

$$v_g = 0.61160c\left[1 + \frac{589.295}{1.63505}\left(-\frac{0.0001}{0.59}\right)\right] = 0.5742c.$$

When we measure the velocity of light with experimental methods using mechanical choppers, we are in fact measuring the group velocity since these methods modulate the light.

8.3 THE DISPERSION RELATION

The dispersion relation for a medium describes how the frequency of a wave ω depends on the wavenumber k. Various dependencies of ω upon k are shown in Figure 8.4. If there is no dispersion a plot of ω against k is a straight line as shown by curve (b), corresponding to:

$$v = \frac{\omega}{k} = \text{constant}, \qquad v_g = \frac{\mathrm{d}\omega}{\mathrm{d}k} = v.$$

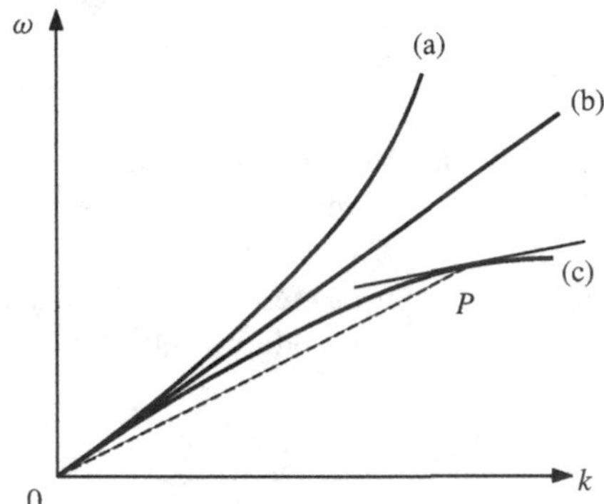

Figure 8.4 Plots of frequency ω against wavenumber k for various dispersion relations, $\omega = \omega(k)$. The straight line, curve (b), corresponds to the non-dispersive case. Curve (a) corresponds to anomalous dispersion while curve (c) corresponds to normal dispersion, where the slope $\mathrm{d}\omega/\mathrm{d}k$ is always less than the gradient ω/k at any point on the curve.

In a dispersive medium a plot of ω against k is nonlinear. For example, curve (c), for a particular dispersive medium, bends 'downwards' as k increases. As illustrated by Figure 8.4, the slope $\mathrm{d}\omega/\mathrm{d}k$ of this curve at any point, e.g. point P, is always less than the gradient ω/k at that point so that the group velocity v_g is always less than the phase velocity v. This is the case of normal dispersion, cf. Equation (8.24). The slope of curve (a), however, bends 'upwards' with increasing k and so v_g is always greater than v. This is the case of anomalous dispersion, cf. Equation (8.24).

We can apply these considerations to the propagation of electromagnetic waves. In vacuum, electromagnetic waves propagate with a velocity

$$v = \frac{1}{\sqrt{\varepsilon_o \mu_o}} = \text{constant}, \tag{8.25}$$

where ε_o and μ_o are the permittivity and permeability of free space, respectively. The velocity, which is the velocity of light, is independent of frequency and the dispersion relation is linear. Consequently, the phase and group velocities are equal. In a *dielectric material* electromagnetic waves travel with a velocity

$$v = \frac{1}{\sqrt{\varepsilon \mu}} \tag{8.26}$$

where ε and μ are the permittivity and permeability of the material, respectively. The refractive index n is given by

$$n = \frac{c}{v} = \sqrt{\frac{\mu\varepsilon}{\mu_o \varepsilon_o}} = \sqrt{\mu_r \varepsilon_r}, \tag{8.27}$$

where $\varepsilon_r = \varepsilon/\varepsilon_o$ and $\mu_r = \mu/\mu_o$ are the relative permittivity and permeability of the material, respectively. For most materials μ_r is constant and approximately equal to 1, but ε_r does vary with frequency giving, $v = \text{constant}/\sqrt{\varepsilon_r}$. We find the group velocity of the electromagnetic waves from Equation (8.24) using

$$\frac{\mathrm{d}v}{\mathrm{d}\lambda} = \frac{\mathrm{d}v}{\mathrm{d}\varepsilon_r}\frac{\mathrm{d}\varepsilon_r}{\mathrm{d}\lambda} = \left(-\frac{1}{2}\frac{v}{\varepsilon_r}\right)\frac{\mathrm{d}\varepsilon_r}{\mathrm{d}\lambda},$$

to obtain

$$v_g = v - \lambda\frac{\mathrm{d}v}{\mathrm{d}\lambda} = v\left(1 + \frac{\lambda}{2\varepsilon_r}\frac{\mathrm{d}\varepsilon_r}{\mathrm{d}\lambda}\right). \tag{8.28}$$

In a medium for which $\mathrm{d}\varepsilon_r/\mathrm{d}\lambda < 0$, it follows that $v_g < v$ and we have normal dispersion. In a medium for which $\mathrm{d}\varepsilon_r/\mathrm{d}\lambda > 0$, $v_g > v$ and we have anomalous dispersion. Dispersion of electromagnetic waves also occurs in the propagation of radio waves in the ionosphere. The ionosphere consists of a gas with some of the molecules ionised by ultraviolet radiation from the sun. Each singly ionised molecule yields a positively charged ion and a free electron. The charged particles affect the velocity of electromagnetic waves that pass through the ionosphere and the resulting dispersion relation is

$$\omega^2 = \omega_o^2 + c^2k^2 \tag{8.29}$$

for frequencies greater than ω_o where ω_o is a constant called the *plasma oscillation frequency*. From Equation (8.29), the phase velocity is given by

$$v = \frac{\omega}{k} = \frac{c}{\sqrt{(1 - \omega_o^2/\omega^2)}}. \tag{8.30}$$

Differentiating Equation (8.29) gives

$$2\omega d\omega = c^2 2k dk.$$

Hence the group velocity is given by

$$v_g = \frac{d\omega}{dk} = c^2 \frac{k}{\omega} = c\sqrt{(1 - \omega_0^2/\omega^2)}. \tag{8.31}$$

Equation (8.30) shows that the phase velocity exceeds the velocity of light c, which appears to violate the special theory of relativity. This theory, however, says that a *signal* cannot propagate at a speed greater than c. Signals travel at the group velocity and Equation (8.31) shows that this is always less than c. We see from Equations (8.30) and (8.31) that

$$v \times v_g = c^2. \tag{8.32}$$

Worked example

When a wave is present on the surface of water there are two types of restoring force that tend to flatten the surface; these forces are gravity and surface tension. The relative strengths of these forces depend upon the wavelength of the waves. For waves on deep water, where the wavelength is small compared with the depth of the water, the angular frequency ω and wavenumber k are related by the dispersion relation

$$\omega^2 = gk + \frac{Sk^3}{\rho},$$

where g is the acceleration due to gravity, and S and ρ are the density and surface tension of water, respectively. Deduce the ratio of the group and phase velocities for (i) the limit of short wavelength and (ii) the limit of long wavelength. At what wavelength are the two velocities equal? (The density and surface tension of water are 1.0×10^3 kg m^{-3} and 7.2×10^{-2} N m^{-1}, respectively; the acceleration due to gravity is 9.81 m s^{-2}.)

Solution

Since

$$\omega = \left(gk + \frac{Sk^3}{\rho}\right)^{1/2}, \tag{8.33}$$

$$v = \frac{\omega}{k} = \left(\frac{g}{k} + \frac{Sk}{\rho}\right)^{1/2}. \tag{8.34}$$

(i) In the limit of short wavelength, $\lambda \to 0$ and $k \to \infty$, and

$$v = \left(\frac{Sk}{\rho}\right)^{1/2} = \frac{\omega}{k}, \text{ giving } \omega = \left(\frac{Sk^3}{\rho}\right)^{1/2}.$$

Hence,

$$v_g = \frac{d\omega}{dk} = \frac{3}{2}\left(\frac{Sk}{\rho}\right)^{1/2} = \frac{3}{2}v.$$

(ii) In the limit of long wavelength, $k \to 0$, and

$$v = \left(\frac{g}{k}\right)^{1/2} \text{ giving } \omega = (gk)^{1/2}.$$

Hence,

$$v_g = \frac{1}{2}\left(\frac{g}{k}\right)^{1/2} = \frac{1}{2}v.$$

To find the wavelength at which the two velocities are equal, we have from Equation (8.34),

$$v_g = \frac{d\omega}{dk} = \frac{1}{2\omega}\left(g + \frac{3Sk^2}{\rho}\right) = \frac{1}{2}\left(gk + \frac{Sk^3}{\rho}\right)^{-1/2}\left(g + \frac{3Sk^2}{\rho}\right).$$

Putting $v_g = v$, using Equation (8.34) for v, and simplifying, we obtain

$$k = \left(\frac{g\rho}{S}\right)^{1/2},$$

giving,

$$\lambda = \frac{2\pi}{k} = 2\pi\left(\frac{7.2 \times 10^{-2}}{9.81 \times 10^3}\right)^{1/2} = 1.7 \times 10^{-2}\text{ m}.$$

For wavelengths much greater than this value, the wave motion is dominated by gravity. For wavelengths much less than this, it is dominated by surface tension.

8.4 WAVE PACKETS

When we superpose i.e. sum two monochromatic waves with nearly equal frequencies we obtain a pattern of beats as shown in Figure 8.1. Of course, we can have a *group* of many waves having different frequencies and in most physical situations this is usually the case. The different frequencies may be discrete or they may cover a continuous range. (We are familiar with the concept of a continuous frequency distribution in the case of white light that contains a continuous range of frequencies from blue to red light.) Figure 8.5(a) illustrates an important example of a continuous frequency distribution that occurs in many physical situations. This distribution lies symmetrically about a central frequency ω_0 and has a width

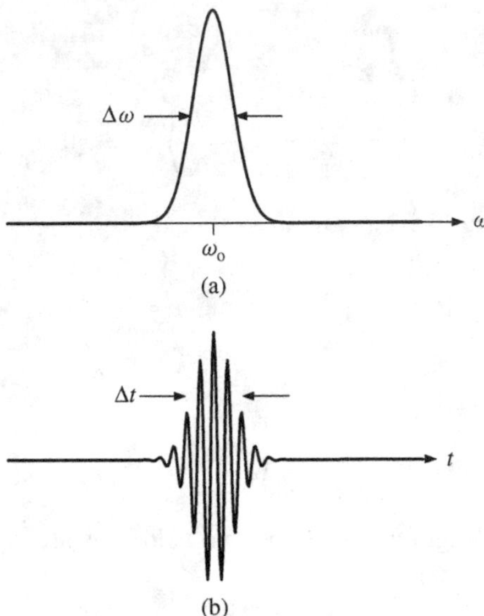

Figure 8.5 (a) An important example of a continuous frequency distribution that occurs in many physical situations. This distribution lies smoothly and symmetrically about a central frequency ω_o. The width $\Delta\omega$ of the distribution is small compared with ω_o. (b) The wave packet, of temporal width Δt, resulting from the superposition of the frequency components of the distribution in (a).

$\Delta\omega$ that is small compared with ω_o. It also has a smooth profile. The result of superposing the frequency components of this distribution is shown on a time axis in Figure 8.5(b). We obtain a pulse of waves or *wave packet* that is highly localised in time with a width Δt. The wave packet travels at the group velocity which is given by the same equation (8.23), $v_g = \mathrm{d}\omega/\mathrm{d}k$, that we had for the case of just two monochromatic waves. The energy is concentrated around the amplitude maximum and travels at the group velocity as does any information carried by the wave packet. In Section 8.4.1 we will show that the width $\Delta\omega$ of the frequency distribution and the temporal width Δt of the wave packet are related by $\Delta t\Delta\omega \approx 2\pi$. This is called the *bandwidth theorem*. This is a very important and general result that applies to a wide range of physical phenomena where there is a disturbance $\psi(t)$ that is localised in time, i.e. some sort of wave pulse. This relationship between Δt and $\Delta\omega$ does not depend on the specific shape of $\psi(t)$ so long as it has the characteristic that defines a pulse, i.e. that $\psi(t)$ is different from zero only over the limited time interval Δt. It follows that to obtain pulses of shorter duration Δt, we have to increase the range of frequencies $\Delta\omega$.

There are many examples of wave pulses and packets in physical situations. For example, narrow pulses of light are passed down optical fibres for communication purposes. Higher data transmission rates require pulses of very short duration Δt. Consequently, the sending and receiving equipment needs to operate over correspondingly high frequency bandwidths. On the research side, scientists are making wave packets of light that contain just a few cycles of optical oscillation, corresponding to pulse lengths of femtoseconds ($\sim 10^{-15}$ s). Wave packets also have special significance in quantum mechanics. There they are interpreted as *probability* waves that describe the position of a particle.

8.4.1 Formation of a wave packet

To illustrate the formation of a wave packet we first consider the superposition of a group of monochromatic waves having a set of *discrete* wavenumbers. Each wave has the form $\psi_n = a_n \cos(k_n x - \omega_n t)$ and their superposition is given by

$$\psi = \sum_n a_n \cos(k_n x - \omega_n t). \tag{8.35}$$

Figure 8.6(b) shows the superposition of a group of eleven such waves and is a snapshot of the resultant wave packet at time $t = 0$. Figure 8.6(a) shows some of the individual waves making up the superposition. [For the sake of clarity only alternate waves are shown and note that Figure 8(a) and (b) have different vertical scales.] These waves have the same amplitude a but their wavenumbers k_n range from $k_o - 5\delta k$ to $k_o + 5\delta k$ in steps δk where $\delta k \ll k_o$. All the individual waves are in phase at $x = 0$ and the amplitude of the superposition at that point is equal

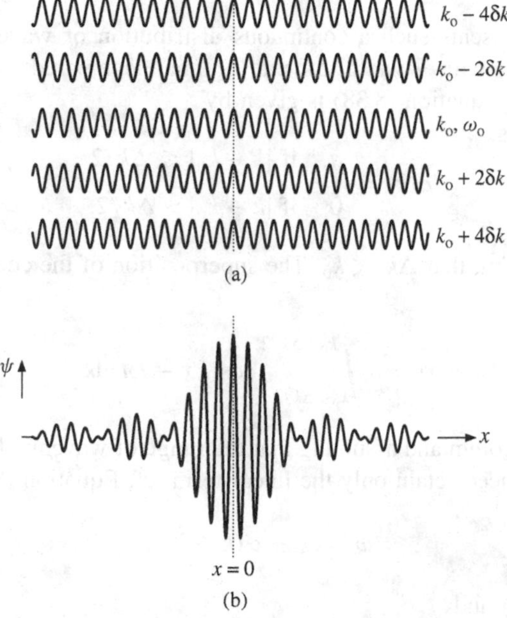

Figure 8.6 (a) Some of the eleven monochromatic waves, contributing to the superposition shown in (b). (Only alternate waves are shown for the sake of clarity.) The eleven waves have the same amplitude but their wavenumbers increase steadily in small steps δk about a mean wavenumber k_o. (b) The wavepacket resulting from the superposition of the eleven waves. The amplitude of the superposition is $11a$ at $x = 0$ when all the eleven waves are in phase with each other. The wavenumber of the wave is equal to the mean k_o of the eleven waves. Note that (a) and (b) have different vertical scales.

to $11a$. As we go away from $x = 0$ in either direction, however, the waves go increasingly out of phase and this leads to a reduction in the amplitude of the superposition, i.e. the formation of a localised wave packet. Equation (8.35) can be recast in the following form:

$$\psi = A(x, t)\cos(k_o x - \omega_o t), \tag{8.36}$$

where

$$A(x, t) = a\frac{\sin[n(x\delta k - t\delta\omega)/2]}{\sin[(x\delta k - t\delta\omega)/2]} \tag{8.37}$$

and n is the number of waves in the group. In analogy to the case of just two monochromatic waves (see Section 8.2.1), the wave travels at the phase velocity ω_o/k_o with a wavenumber equal to the mean k_o of the eleven monochromatic waves while the wave packet travels at the group velocity $\mathrm{d}\omega/\mathrm{d}k$.

Suppose now that we have a group of waves that have a *continuous* distribution of wavenumbers. Then, the summation of Equation (8.35) is replaced by an integral of the form

$$\psi = \int a(k)\cos(kx - \omega t)\mathrm{d}k. \tag{8.38}$$

Figure 8.7(a) represents such a continuous distribution of wavenumbers, centred at wavenumber k_o with a width Δk that is small compared with k_o. The wave amplitude $a(k)$ in Equation (8.38) is given by

$$a(k) = \begin{matrix} a, & \text{if } |k - k_o| \leqslant \Delta k/2 \\ 0, & \text{if } |k - k_o| > \Delta k/2, \end{matrix}$$

and we are assuming that $\Delta k \ll k_o$. The superposition of the corresponding group of waves is

$$\psi = a\int_{k_o-\Delta k/2}^{k_o+\Delta k/2} \cos(kx - \omega t)\mathrm{d}k. \tag{8.39}$$

Using Taylor's theorem and assuming that the range of wavenumbers is sufficiently small so that we need retain only the linear term, cf. Equation (8.19), we have

$$\omega = \omega_o + \alpha(k - k_o), \tag{8.40}$$

where $\omega_o = \omega(k_o)$ and

$$\alpha \equiv \left(\frac{\mathrm{d}\omega}{\mathrm{d}k}\right)_{k=k_o}. \tag{8.41}$$

Hence, substituting Equation (8.40) for ω in $(kx - \omega t)$:

$$kx - \omega t = kx - [\omega_o + \alpha(k - k_o)]t = k(x - \alpha t) - \beta t$$

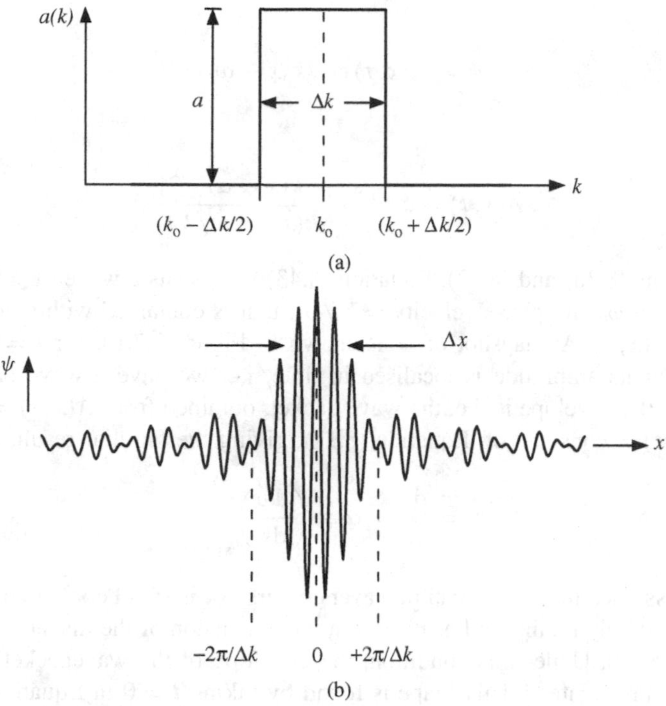

Figure 8.7 (a) A continuous distribution of wavenumbers, centred at wavenumber k_o with a width Δk that is small compared with k_o and a constant amplitude a. (b) The wave packet that results from the superposition of the continuous distribution in (a). The width of the wave packet is taken to be equal to $2\pi/\Delta k$.

where $\beta \equiv \omega_o - \alpha k_o$. We introduce $\xi = k(x - \alpha t) - \beta t$ as the new variable of integration. Hence

$$d\xi = (x - \alpha t)dk,$$

and we can rewrite Equation (8.39) as

$$\psi = a \int_{\xi_1}^{\xi_2} \frac{\cos \xi \, d\xi}{(x - \alpha t)}$$

with the range of integration from $\xi_1 = (k_o - \Delta k/2)(x - \alpha t) - \beta t$ to $\xi_2 = (k_o + \Delta k/2)(x - \alpha t) - \beta t$. Hence,

$$\psi = \frac{a}{(x - \alpha t)}(\sin \xi_2 - \sin \xi_1).$$

Using the trigonometric identity

$$\sin \xi_2 - \sin \xi_1 = 2 \sin [(\xi_2 - \xi_1)/2] \cos [(\xi_2 + \xi_1)/2], \tag{8.42}$$

we obtain

$$\psi = A(x,t)\cos(k_o x - \omega_o t) \tag{8.43}$$

where

$$A(x,t) = a\Delta k \frac{\sin[\Delta k(x - \alpha t)/2]}{\Delta k(x - \alpha t)/2}, \tag{8.44}$$

cf. Equations (8.36) and (8.37). Equation (8.43) represents a wave of wavenumber k_o, frequency ω_o and phase velocity $= k_o/\omega_o$, that is contained within an envelope given by $A(x,t)$. A snapshot of ψ is shown in Figure 8.7(b), for $t = 0$, and we can see that its amplitude is localised in time, i.e. we have a wave packet. The velocity of the envelope is, i.e. the wave packet, obtained from $A(x,t) =$ constant, i.e. $(x - \alpha t) =$ constant, cf. Equation (8.17), giving the familiar result,

$$v_g = \frac{\mathrm{d}x}{\mathrm{d}t} = \alpha \equiv \left(\frac{\mathrm{d}\omega}{\mathrm{d}k}\right)_{k=k_o}.$$

We have assumed that the spread in wavenumbers Δk is small compared with k_o so that we need only retain the linear term in the expansion of the dispersion relation, Equation (8.40). Under this condition, the envelope of the wavepacket retains its shape as it propagates.[2] This shape is found by taking $t = 0$ in Equation (8.44):

$$A(x) = a\Delta k \frac{\sin(x\Delta k/2)}{x\Delta k/2}. \tag{8.45}$$

The function $[\sin(x\Delta k/2)]/(x \sin \Delta k/2)$ is the now familiar sinc function. At $x = 0$, it has the value unity. It first becomes equal to zero when $x\Delta k/2 = \pm\pi$, giving

$$x = \pm\frac{2\pi}{\Delta k}. \tag{8.46}$$

Δk is the width of the wavenumber distribution, see Figure 8.7(a). For a measure of the width Δx of the wave packet we could chose the distance between the first two zeros of $A(t)$, i.e. the first two zeros of the sinc function. However, in practice it is more usual to take one half of this value, i.e. $\Delta x = 2\pi/\Delta k$. We thus find that the product of the wavenumber spread Δk and the width Δx of the resultant wave packet is given by

$$\boxed{\Delta x \Delta k \approx 2\pi} \tag{8.47}$$

where the symbol $\approx$ indicates the imprecision in the measure of the wave packet width. This is an example of the bandwidth theorem that we introduced in Section 8.4. Here it says that the shorter the length of the wave packet, the greater

[2] If this condition does not hold, we must retain higher terms in $(k - k_o)$ in the expansion of the dispersion relation, and the shape of the envelope will change as it propagates.

is the range of wavenumbers that is necessary to represent it. For a monochromatic wave Δk is zero and so the wave is infinitely long, as we have noted before. We can also express the bandwidth theorem in terms of frequency and time. A wave packet that is Δx long and travels at velocity $v_g = \mathrm{d}\omega/\mathrm{d}k$ takes time Δt to pass a fixed point where

$$\Delta t = \frac{\Delta x}{\mathrm{d}\omega/\mathrm{d}k}. \tag{8.48}$$

Hence we can write

$$\Delta t \Delta\omega = \Delta x \Delta k, \tag{8.49}$$

and so

$$\boxed{\Delta t \Delta\omega \approx 2\pi.} \tag{8.50}$$

This is the relationship given in Section 8.4. The bandwidth theorem expresses the fact that a wave packet (or pulse) of duration Δt is the superposition of frequency components over the range $\Delta\omega$ and the shorter the duration of the wave packet, the wider the range of frequencies required to represent it, cf. Figure 8.5. These concepts are closely related to the Heisenberg *Uncertainty Principle* in quantum wave mechanics where particles are described in terms of waves. The position of a particle in the one-dimensional case is defined as 'somewhere' within a wave group of length Δx. The wavelength λ of a particle is related to its momentum p by de Broglie's relationship

$$\lambda = \frac{h}{p}, \tag{8.51}$$

where h is Planck's constant. From Equation (8.47) and using $\lambda = 2\pi/k$, it readily follows that

$$\Delta x \Delta p \approx h. \tag{8.52}$$

This is an expression of the Uncertainty Principle. It says that the wave nature of a particle makes it impossible to know, at the same time, both its position and momentum beyond the condition imposed by Equation (8.52).

PROBLEMS 8

8.1 Two laser beams that have nearly the same wavelength can produce a beat frequency if they are incident on a photodetector with a sufficiently fast response time. One laser has a wavelength of 766.49110 nm while a second laser has a slightly shorter wavelength. They produce a beat frequency of 462 MHz. What is the wavelength of the second laser?

8.2 The A and E strings of a violin are tuned to frequencies of 440 Hz and 660 Hz, respectively. A musician finds that the E string on her violin is sharp. By playing the A and E strings simultaneously she hears a beat frequency of 4 Hz. (a) Why do the beats arise? (b) What is the actual frequency of the E string?

8.3 The velocity of a wave pulse on a taut string can be determined by measuring the time it takes the pulse to travel the distance between the two fixed ends. Alternatively, the velocity of a wave on the string can be determined from a measurement of the frequency of the fundamental mode of the vibrating string. Does each of these methods determine phase velocity or group velocity?

8.4 (a) Find the relationship between the group velocity v_g and the phase velocity v for (i) a medium for which v is inversely proportional to wavelength λ and (ii) a medium for which v is proportional to $(\lambda)^{-1/2}$. (b) The dispersion relation for electromagnetic waves in vacuum is $\omega = ck$, where c is the velocity of light. Determine the phase and group velocities of such waves, showing that they are equal. (c) The relative permittivity ε_r of an ionised gas is given by

$$\varepsilon_r = \frac{c^2}{v^2} = 1 - \frac{\omega_o^2}{\omega^2},$$

where ω_o is the plasma oscillation frequency. Show that this leads to the dispersion relation

$$\omega^2 = \omega_o^2 + c^2k^2.$$

8.5 (a) Calculate (i) the phase velocity and (ii) the group velocity for deep ocean waves at a wavelength of 100 m. (b) Determine the minimum value of the phase velocity of water waves on deep water. (The density and surface tension of water are 1.0×10^3 kg m^{-3} and 7.2×10^{-2} N m^{-1}, respectively; assume $g = 9.81$ m s^{-2}.)

8.6 A rectangular dish containing mercury is connected to the cone of a loudspeaker so that when the loudspeaker is driven by an oscillating voltage a standing wave is set up on the surface of the mercury. When a beam of light is shone on the surface, the standing wave acts like a *diffraction grating* and the observed diffraction pattern enables the spacings of the antinodes of the standing wave to be determined. It is found that the spacing of the antinodes for a standing wave of frequency 1.35 kHz is 0.25 mm. (a) Use these data to obtain a value for the surface tension S of mercury. Assume the dispersion relation

$$\omega^2 = gk + \frac{Sk^3}{\rho},$$

where ρ is the density and assume also that the wavelength is sufficiently small that the wave properties are determined by surface tension and not by gravity. (b) What is the value of the group velocity?
(The density of mercury = 13.6×10^3 kg m^{-3}; assume $g = 9.81$ m s^{-2}.)

8.7 Cauchy's formula is an empirical relationship that relates the refractive index n of a transparent medium to wavelength λ, where λ is the wavelength of the light in vacuum. The formula is $n = A + B/\lambda^2$, where A and B are constants for the particular medium. (a) Show that the ratio of group and phase velocities at wavelength λ is given by

$$\frac{v_g}{v} = \frac{(A - B/\lambda^2)}{(A + B/\lambda^2)}.$$

(b) Evaluate this ratio at a wavelength of 600 nm for a particular type of glass for which $A = 1.45$ and $B = 3.6 \times 10^{-14}$ m^2.

8.8 When a transverse wave travels down a real wire there are forces acting on each portion of the wire in addition to the force resulting from the tension in the wire. An equation that gives an improved description of a wave on a real wire is

$$\frac{\partial^2 y}{\partial t^2} = \frac{T}{\mu}\left(\frac{\partial^2 y}{\partial x^2}\right) - \alpha y,$$

where T is the tension, μ is the mass per unit length and α is a constant. (a) Show that $y = A\cos(\omega t - kx)$ is a solution to this equation subject to the condition

$$\omega^2 = \frac{T}{\mu}k^2 + \alpha.$$

(b) What is the lowest angular frequency that the wire can support according to this condition? (c) Obtain the relationship between the group and phase velocities for waves on the wire.

8.9 An amplifier is used to increase the amplitude of a voltage pulse that has a temporal width of 5×10^{-8} s. Estimate the required frequency bandwidth (in Hz) of the amplifier.

8.10 A *free electron laser* is a device that can produce a very short pulse of light. If the width of the light pulse is 100 fs ($= 100 \times 10^{-15}$ s) and the central wavelength of the pulse is 500 nm, estimate the spread of wavelengths in the light pulse.

8.11 A group of n monochromatic waves of equal amplitude a have wavenumbers that span the range Δk in steps δk. The superposition of these waves is given by Equations (8.36) and (8.37):

$$\psi = A(x,t)\cos(k_o x - \omega_o t), \text{ where } A(x,t) = a\frac{\sin[n(x\delta k - t\delta\omega)/2]}{\sin[(x\delta k - t\delta\omega)/2]}.$$

(a) Obtain an expression for $A(x,t)$ for the case where $n = 2$. (b) Consider the situation where n becomes very large but the product $(n-1)\delta k = \Delta k$ remains constant. Show that for this case, we can write

$$A(x,t) = na\frac{\sin[\Delta k(x - \alpha t)/2]}{\Delta k(x - \alpha t)/2},$$

where $\alpha = \delta\omega/\delta k$. Compare the expressions from (a) and (b) with Equations (8.15b) and (8.44), respectively.

Appendix: Solutions to Problems

SOLUTIONS 1

1.1 (a) (i) 4.0 s, (ii) $\pi/2$ rad s^{-1}, (iii) 1.23 N m^{-1}.
1.2 (a) 1.38 m s^{-1}, (b) 3.82×10^3 m s^{-2}.
1.3 $a_{max} < g$, giving $\nu_{max} = 1.1$ Hz.
1.4 (a) Potential energy is 25% of total energy and hence kinetic energy is 75% of total energy. (b) (i) Total energy is quadrupled, (ii) maximum velocity is doubled and (iii) maximum acceleration is doubled.
1.5 (a) 0.41 J. (b) $x = 0.045\cos(23t + 2.7)$m.
1.6 For the system of two springs connected in parallel, the force on the mass is the sum of the forces due to the separate springs, giving $\omega_a = \sqrt{2k/m} = \sqrt{2}\omega_b$. For the system of two springs connected in parallel the tension in both springs must be the same, giving $\omega_c = \sqrt{k/2m} = \omega_b/\sqrt{2}$.
1.7 (a) When the test tube is displaced a distance x into the liquid, the restoring force due to buoyancy is $-A\rho g x$. Hence, equation of motion is

$$m\frac{d^2x}{dt^2} = -A\rho g x.$$

This is SHM with frequency $\omega = \sqrt{A\rho g/m}$.

(b) $F = -A\rho g x$, giving $U = \int_0^x A\rho g x' dx' = \frac{1}{2}A\rho g x^2$.

Hence, $E = \frac{1}{2}mv^2 + \frac{1}{2}A\rho g x^2$, where v is the velocity of the test tube.

1.8 We denote the fundamental quantities mass, length and time by M, L and T, respectively. Since the dimensions of g are L T^{-2} we have,

$$T \equiv M^{\alpha}L^{\beta}[LT^{-2}]^{\gamma}.$$

The dimensions of both sides of this equation must be the same and equating indices of M, L and T we obtain

$$\alpha = 0, \beta + \gamma = 0, -2\gamma = 1,$$

giving, $T \propto \sqrt{l/g}$.

1.9 Starting from Equation (1.36), obtain: (a) 1.81×10^{-2} m s^{-1}, (b) 0.43 s.

1.10 Couple acting on rod $= -kL\sin\theta \times L\cos\theta = -kL^2\theta$ for small θ. Hence, $I\dfrac{\mathrm{d}^2\theta}{\mathrm{d}t^2} = -kL^2\theta$.

1.11 (a)
$$F = -\frac{\mathrm{d}U}{\mathrm{d}x} = -\frac{6a}{x^7} + \frac{12b}{x^{13}}.$$
At equilibrium, $F = 0$, giving $x_o = (2b/a)^{1/6}$.

(b) For displacement Δx from equilibrium, Taylor's theorem gives

$$F(x_o + \Delta x) = F(x_o) + \Delta x\left(\frac{\mathrm{d}F}{\mathrm{d}x}\right)_{x=x_o} + \cdots.$$

$$F(x_o) = 0, \text{ and } \left(\frac{\mathrm{d}F}{\mathrm{d}x}\right)_{x=x_o} = \frac{42a}{x_o^8} - \frac{156b}{x_o^{14}}.$$

Hence, neglecting higher terms,

$$F(x_o + \Delta x) = -36a(a/2b)^{4/3}\Delta x.$$

This gives SHM with frequency $\sqrt{k/m}$ where m is reduced mass and $k = -36a(a/2b)^{4/3}$.

1.12 (a) Consider an elemental length $\mathrm{d}l$ of spring at a distance l from the support. Mass of element $= m\mathrm{d}l/l_o$ where l_o is the equilibrium length of the spring. Velocity of element $= vl/l_o$.

Hence kinetic energy of spring $= \dfrac{1}{2}\dfrac{mv^2}{l_o^3}\displaystyle\int_0^{l_o} l^2\mathrm{d}l = \dfrac{1}{6}mv^2$.

Kinetic energy of mass $M = 1/2Mv^2$ and potential energy of extended spring $= 1/2kx^2$. Hence the total energy of the system (i.e. of spring plus mass M) is

$$E = \frac{1}{2}(M + m/3)v^2 + \frac{1}{2}kx^2.$$

(b) Since the total energy E of the system is conserved, $\dfrac{\mathrm{d}E}{\mathrm{d}t} = 0$, from which it follows that $(M + m/3)\dfrac{\mathrm{d}v}{\mathrm{d}t} = -kx$. This is SHM with $\omega = \sqrt{k/(M + m/3)}$.

1.13 (a) From conservation of energy,

$$\frac{1}{2}mv^2 + U(x) = \text{constant} = U(A) \text{ and hence,}$$

$v = \sqrt{2[U(A) - U(x)]/m}$.

(b) From $v = \mathrm{d}x/\mathrm{d}t$, $\mathrm{d}t = \mathrm{d}x/v$ and hence

$$\text{period } T = \int_{\text{period}} \mathrm{d}t = 2\int_{-A}^{A} \frac{\mathrm{d}x}{v} = 4\int_{0}^{A} \frac{\mathrm{d}x}{v} \text{ for symmetric potential}$$

$$= 4\sqrt{\frac{m}{2U(A)}}\int_0^A \frac{\mathrm{d}x}{\sqrt{(1 - U(x)/U(A))}}.$$

(c) For $U(x) = \alpha x^n$ and letting $\xi = x/A$, obtain

$$T = 4\sqrt{\frac{m}{2\alpha A^n}}\int_0^1 \frac{A\mathrm{d}\xi}{\sqrt{[1-\xi^n]}}$$

$$= \frac{1}{A^{(n/2)-1}} \times (\text{factor independent of } A).$$

Hence, for $n = 2 : T$ is independent of amplitude A,
for $n = 4 : T \propto 1/A$, etc.

SOLUTIONS 2

2.1 We require the condition of critical damping for which $b/2m = \omega_o = \sqrt{k/m}$.
This gives $b = 2m\sqrt{g/\text{spring extension}}$.
Hence $b = 64 \text{ kg s}^{-1}$.

2.2 Using Equation (2.9) and $\gamma = b/m$, obtain
$b = (2m/T)\ln(1/0.90) = 0.042 \text{ kg s}^{-1}$, taking $T = 2.5$ s.
Hence, damping force $= -0.042v$ N, where v is the velocity.
$\gamma = 0.084 \text{ s}^{-1}$.

2.3 Using $A(t) = A_0 \exp(-\gamma t/2)$ and $Q = \omega_o/\gamma$, obtain $Q = \dfrac{\omega_o t}{2\ln[A_0/A(t)]}$.

Inspection of graph shows 20 complete cycles of oscillation take 600 s and amplitude falls by a factor of approximately 2.8 during this time. Using $\omega_o = 2\pi/T$ and with $t = 20T$,

$$Q = \frac{2\pi \times 20T}{2T\ln(2.8)} \approx 60.$$

2.4 Using Equation (2.18):

$$E(t = 10T) = E_o \exp(-10\gamma T) = E_o/2, \text{ giving } \exp(-10\gamma T) = 1/2.$$

$$E(t = 50T) = E_o \exp(-50\gamma T) = E_o \exp(-10\gamma T)^5 = E_o(1/2)^5.$$

Hence, energy after 50 cycles is reduced by a factor of 32.

2.5 (a) Q-values: 314, 10.5 and 3.14, respectively.
ω-values: 3.142, 3.138 and 3.102, respectively, which do not change appreciably.

(c) Using

$$x = A\exp(-\gamma t/2) + Bt\exp(-\gamma t/2)$$

and

$$\frac{dx}{dt} = \exp(-\gamma t/2)[B - \gamma Bt/2 - \gamma A/2]$$

with initial conditions at $t = 0 : x = 10$ and $dx/dt = 0$, obtain

$$x = 10\exp(-\pi t)(1 + \pi t), \text{ since } \gamma = 2\omega_o = 2\pi.$$

2.6 Use Equations (2.6) and (2.21) to obtain $\omega = \omega_o(1 - 1/4Q^2)^{1/2}$ and the approximation $(1 - \alpha)^{1/2} \simeq 1 - \alpha/2$ for $\alpha \ll 1$.

2.7 For pendulums we have

$$m\frac{d^2x}{dt^2} + b\frac{dx}{dt} + \frac{mg}{l} = 0, \text{ where } b \text{ is a constant.}$$

Since $A(t) = A_0\exp(-b/2m)t$,

$$\frac{\ln[A(t)/A_0]_{\text{brass}}}{\ln[A(t)/A_0]_{\text{alum.}}} = \frac{\rho_{\text{alum.}}}{\rho_{\text{brass}}}, \text{ where } \rho \text{ is the density.}$$

Hence, $\ln[A(t)/A_0]_{\text{brass}} = \ln(0.5)\frac{2.7}{8.5}$, and the amplitude of the brass pendulum is reduced by a factor of 0.80.

2.8 (a) Energy loss per cycle $= \frac{Ke^2A^2\omega^4}{c^3}\int_0^T \sin^2\omega t\,dt$.

(b) Use $\frac{\text{energy loss per cycle}}{\text{stored energy}} = \frac{2\pi}{Q}$.

(c) $\tau \equiv 1/\gamma = Q/\omega = mc\lambda^2/Ke^2 4\pi^2$. For $\lambda = 500$ nm, $\tau \approx 1 \times 10^{-8}$ s.

SOLUTIONS 3

3.1 Use Equations (3.18) and (3.12).
(a) 0.013 m, 0.58°, (b) 0.13 m, 90°, (c) 5.2×10^{-4} m, 179°.

3.2 Follow the hints to obtain

$$A = \frac{a}{(1 + 1/u^4 - 2/u^2 + 1/u^2Q^2)^{1/2}}.$$

For A to be a maximum, the denominator must be a minimum.

3.3 (a) $\frac{\omega_o - \omega_{\max}}{\omega_o} = 1 - \left(1 - \frac{1}{2Q^2}\right)^{1/2} \simeq \frac{1}{4Q^2}$ for $2Q^2 \ll 1$. Answer: 0.25%.

(b) Similarly, $\frac{A_{\max} - A(\omega_o)}{A(\omega_o)} \simeq \frac{1}{8Q^2}$ for $4Q^2 \ll 1$. Answer: 0.125%.

3.4 Close to the resonance frequency and for the given parameters,

$$\overline{P}(\omega) = \frac{50}{[(\omega - 100)^2/4] + 1}\,\text{W}.$$

3.5 (a) 398 Hz.
(b) At resonance frequency, impedance of circuit $= R$ giving $I_0 = 0.2$ A.

3.6 $e^{i\pi/2} = i$ and hence $i^i = e^{-\pi/2} = 0.208$.

3.7 $dz/dt = i\omega z$, where the factor i implies a phase difference of $\pi/2$ between z and dz/dt and indeed between x and dx/dt. The sign of the phase shift shows that dx/dt is in advance of x.

3.8 (a) When the pendulum mass is at a distance x from its equilibrium position and the point of suspension is at a distance ξ from its equilibrium position, the restoring force on the mass is

$$-mg\sin[(x-\xi)/l] = -mg(x - a\cos\omega t)/l.$$

Using the small-angle approximation, this leads to the equation of motion:

$$m\frac{d^2x}{dt^2} + b\frac{dx}{dt} + m\omega_o^2 x = m\omega_o^2 a\cos\omega t$$

which is the real part of the complex equation

$$m\frac{d^2z}{dt^2} + b\frac{dz}{dt} + m\omega_o^2 z = m\omega_o^2 a e^{i\omega t}.$$

3.9 (a) Using $A(t) = A_0 e^{-\gamma t/2}$ and $Q = \omega_o/\gamma$ obtain $Q = \dfrac{n\pi}{\ln[A_0/A(t)]}$ where n is the number of complete cycles in time t.
Hence $Q = 75\pi$.
(b) Resonance amplitude $\simeq Qa = 0.12$ m.
(c) Starting with:

$$A(\omega) = a\frac{\omega_o^2}{[(\omega_o^2 - \omega^2)^2 + \omega^2\gamma^2]^{1/2}},$$

half height points will occur at frequencies where

$$[(\omega_o^2 - \omega^2)^2 + \omega^2\gamma^2]^{1/2} = 2[\omega_o^2\gamma^2]^{1/2}.$$

Hence,

$$[(\omega_o - \omega)(\omega_o + \omega)]^2 + \omega^2\gamma^2 = 4\omega_o^2\gamma^2.$$

Letting $\omega_o - \omega = \Delta\omega$ and making the approximation that $\omega = \omega_o$ near to the resonance frequency, obtain $\Delta\omega = \gamma\sqrt{3}/2$ and hence resonance width $= 2\gamma = \gamma\sqrt{3} = \dfrac{\sqrt{3}}{Q}\sqrt{\dfrac{g}{l}}$ which is equal to 0.019 rad s^{-1}.

3.10 (a) (iii)

$$E = \frac{1}{2}mA^2\omega^2 \sin^2(\omega t - \delta) + \frac{1}{2}kA^2 \cos^2(\omega t - \delta)$$
$$= \frac{1}{2}mA^2[\omega^2 \sin^2(\omega t - \delta) + \omega_o^2 \cos^2(\omega t - \delta)].$$

(b) Differentiate E with respect to t and equate the result to zero to obtain $\omega = \omega_o$ when $E = \frac{1}{2}mA^2\omega_o^2$ where A is the amplitude at resonance.

(c)
$$\frac{\overline{K}}{\overline{E}} = \frac{1}{1 + (\omega_o/\omega)^2}.$$

(d)
$$\overline{E} = \overline{K} + \overline{U} = \frac{1}{4}mA^2(\omega_o^2 + \omega^2).$$

Then, substitute for $A = \dfrac{F_0/m}{[(\omega_o^2 - \omega^2)^2 + \omega^2 b^2/m^2]^{1/2}}$, cf. Equation (3.18).

3.11 (a) Energy loss/cycle $= bv_o^2 \int_0^T \sin^2(\omega t - \delta)\mathrm{d}t = \dfrac{bv_o^2 T}{2} = \pi b A^2 \omega.$

(b) Recall that energy of a simple harmonic oscillator $= \frac{1}{2}m\omega^2 A^2$.

(c) Take $\omega = \omega_o$ at resonance.

3.12 Total energy dissipated $= Mgh$, where M is mass of winding weight and h is the distance it falls in 8 days. Total number of cycles $= T'/T$ where $T' = 8$ days and $T = 2\pi\sqrt{l/g}$ is the period of the pendulum. Stored energy $= \frac{1}{2}mgA^2/l$.

$$\text{Using } \frac{\text{energy dissipated/cycle}}{\text{stored energy}} = \frac{2\pi}{Q}, \text{ obtain } Q = \frac{\pi m A^2 T'}{MlhT} \simeq 70.$$

SOLUTIONS 4

4.1 (a) $\omega_1 = 5.72$ rad s^{-1} and $\omega_2 = 5.99$ rad s^{-1}.

(b) Using $x_a = A \cos \dfrac{(\omega_2 - \omega_1)t}{2} \cos \dfrac{(\omega_2 + \omega_1)t}{2}$, we have a high frequency oscillation whose amplitude is modulated at the lower frequency $(\omega_2 - \omega_1)/2$. Amplitude becomes zero after one quarter of the lower frequency, cf. Figure 4.8, $= \dfrac{1}{4}\dfrac{2\pi}{(\omega_2 - \omega_1)/2} = 11.6$ s.

4.2 (a) At time $t = 0$, $x_a = 1/2(C_1 + C_2)$, $x_b = 1/2(C_1 - C_2)$. Hence, $C_1 = 10$ mm, $C_2 = 0$ mm.

(b) $C_1 = 0$ mm, $C_2 = 10$ mm, (ii) $C_1 = 10$ mm, $C_2 = 10$ mm, (iii) $C_1 = 15$ mm, $C_2 = 5$ mm.

4.3 $q_1 = (x_a + x_b)$ with $\omega_1 = \sqrt{k/m}$, $q_2 = (x_b - x_a)$ with $\omega_2 = \sqrt{3k/m}$.

4.4 Energy of a simple harmonic oscillator $= (1/2)m\omega^2(\text{amplitude})^2$. In this case $\omega = (\omega_2 + \omega_1)/2$ and amplitude $= A\cos[(\omega_2 - \omega_1)t]/2$ or $A\sin[(\omega_2 - \omega_1)t]/2$. Frequency of exchange of energy $= (\omega_2 - \omega_1)$.

4.5 (a) Tensions in upper and lower strings are $2mg$ and mg, respectively, and are assumed to be constant during oscillations.
Take $\sin\theta_1 = x_1/l$ and $\sin\theta_2 = (x_2 - x_1)/l$.
Then, $m\dfrac{\mathrm{d}^2x_1}{\mathrm{d}t^2} = -\dfrac{2mg}{l}x_1 + \dfrac{mg}{l}(x_2 - x_1)$, etc.

(b) $B/A = 1 \pm \sqrt{2}$ for $\omega = \sqrt{(2 \pm \sqrt{2})g/l}$, respectively.

(c) 1.1 s, 2.6 s and 2.0 s, respectively.

4.6 (a) The centre of mass of the system remains stationary during the vibrations. In the symmetric-stretch mode the central mass also remains stationary. The other two masses vibrate against the central mass (moving in opposite directions) at the same frequency which is that of a mass m on a spring of spring constant k, i.e. $\sqrt{k/m}$.

(b) The tensions in the left-hand and right-hand springs are $T_1 = k(x_2 - x_1)$ and $T_2 = k(x_3 - x_2)$, respectively. This leads to the stated equations of motion.

(c) Assuming solutions of the form of normal coordinates, i.e. $x_1 = A\cos\omega t$, $x_2 = B\cos\omega t$ and $x_3 = C\cos\omega t$, the equations of motion lead to $A(\omega_1^2 - \omega^2) = C(\omega_1^2 - \omega^2)$.
The solutions for ω of this equation give the normal frequencies:

(i) $\omega = \omega_1 = \sqrt{k/m}$, the first normal mode frequency;

(ii) $A = C$ gives the second normal mode frequency

$$\omega_2 = \sqrt{k(2m + M)/Mn}.$$

(d) Ratio $\dfrac{\omega_2}{\omega_1} = \sqrt{\dfrac{2m}{M} + 1} = 1.91$ which compares with the value of $7/4 = 1.75$ from the text.

4.7 (a) Letting the downward displacements of the upper and lower masses be x_1 and x_2, respectively, the tensions in the upper and lower springs are $4kx_1$ and $k(x_2 - x_1)$, respectively. This leads to the following equations of motion:

$$3m\frac{\mathrm{d}^2x_1}{\mathrm{d}t^2} + 5kx_1 - kx_2 = 0; \qquad m\frac{\mathrm{d}^2x_2}{\mathrm{d}t^2} - kx_1 + kx_2 = 0.$$

Assuming solutions of the form, $x_1 = A\mathrm{e}^{i\omega t}$, $x_2 = B\mathrm{e}^{i\omega t}$ and solving resulting equations for ω gives normal frequencies $\sqrt{2k/m}$ and $\sqrt{2k/3m}$.

(b) For $\omega = \sqrt{2k/m}$, $B = -A$. This means that at any instance, the masses are equidistant from their equilibrium positions and are on opposite sides of them. For $\omega = \sqrt{2k/3m}$, $B = 3A$. At any instance, the masses are both either above or below their equilibrium positions, the displacement of the lower mass being three times that of the higher mass.

4.8 There are five normal modes, as illustrated.

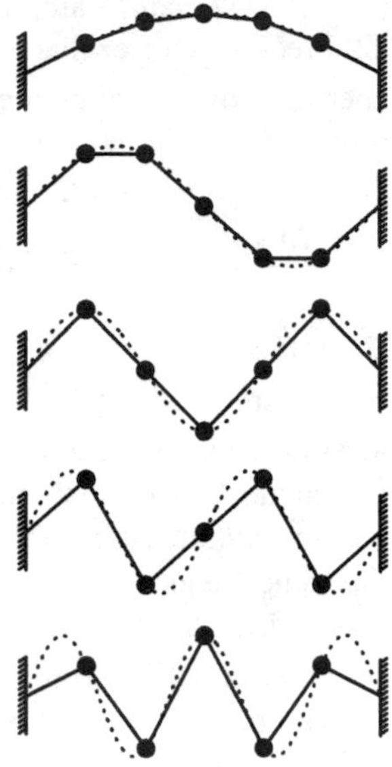

4.9 (a) For downward displacements x_1 of mass M and x_2 of mass m, tensions in the top and bottom springs are k_1x_1 and $k_2(x_2 - x_1)$, respectively, leading to the stated equations of motion.

(c) Substituting $\omega = \sqrt{k_1/M}$ in equation for B gives the desired result. $\omega = \sqrt{k_1/M}$ is, of course, the resonance frequency of a mass M connected to a spring of spring constant k_1.

4.10 (a) From left to right, the tensions in the springs are $T_1 = kx_1, T_2 = k(x_2 - x_1), T_3 = k(x_3 - x_2)$, etc.
This leads to equations of motion,

$$m\frac{\mathrm{d}^2x_1}{\mathrm{d}t^2} = -2kx_1 + kx_2, \text{ etc.}$$

Use of usual substitutions, e.g., $x_1 = A\cos\omega t, x_2 = B\cos\omega t$ and $x_3 = C\cos\omega t$, leads to:

$$\begin{bmatrix} (2\omega_0^2 - \omega^2), & -\omega_0^2, & 0 \\ -\omega_0^2, & (2\omega_0^2 - \omega^2), & -\omega_0^2 \\ 0, & -\omega_0^2, & (2\omega_0^2 - \omega^2) \end{bmatrix} \begin{bmatrix} A \\ B \\ C \end{bmatrix} = 0$$

where $\omega_0 = \sqrt{k/m}$. For non-zero solutions we require the determinant to vanish, giving $(2\omega_0^2 - \omega^2)(\omega^4 - 4\omega_0^2\omega^2 + 2\omega_0^2) = 0$. Solving this equation for ω gives the normal frequencies $\sqrt{2k/m}$ and $\sqrt{(2 \pm \sqrt{2})k/m}$.

(b) For $\omega = \sqrt{2k/m}$, $A = -C$, $B = 0$.

For $\omega = \sqrt{(2+\sqrt{2})k/m}$, $A = C$, $B = -\sqrt{2}A$.

For $\omega = \sqrt{(2-\sqrt{2})k/m}$, $A = C$, $B = \sqrt{2}A$.

SOLUTIONS 5

5.1 Amplitude = 15 mm, wavelength = 8π mm, frequency = 11.9 Hz and velocity = 300 mm s^{-1}. The wave travels in the negative x-direction.

5.2 Amplitude $A = 0.15$ m, $\omega = 20\pi$ rad s^{-1}, $\lambda = 5.0$ m, and $k = 2\pi/5$. Cosine solution is the appropriate one, since displacement $= A$ at $x = 0$, $t = 0$. Wave travels in the positive x-direction. Hence equation is $y = 0.15\cos(0.4\pi x - 20\pi t)$ m.

5.3 (a) (iii) Make use of $T = \lambda/v$. (iv) Make use of $k = 2\pi/\lambda$, $\omega = 2\pi\nu$ and $\lambda\nu = v$.

(b) $v = \dfrac{\omega}{(k_1^2 + k_2^2 + k_3^2)^{1/2}}$.

5.4 Make use of trignometric relations

$$\cos(\alpha - \beta) = \cos\alpha\cos\beta + \sin\alpha\sin\beta$$

$$\sin(\alpha - \beta) = \sin\alpha\cos\beta - \cos\alpha\sin\beta.$$

(a) $A\cos(\omega t - kx) = A\cos(kx - \omega t)$, i.e. no difference between the waves they describe.

(b) $A\sin(\omega t - kx) = -A\sin(kx - \omega t) = A\sin(\omega t - kx \pm \pi)$, i.e. a phase difference of $\pm\pi$ between the waves.

5.5

$$y(x + \delta x, t + \delta t) = A\exp\left\{-\frac{[x + \delta x - v(t + \delta t)]^2}{a^2}\right\}$$

$$= A\exp\left\{-\frac{[x - vt]^2}{a^2}\right\}, \text{ since } v\delta t = \delta x.$$

5.6 (a) (i) 2.0×10^5 Hz, (ii) 6.0×10^{14} Hz, (iii) 3.0×10^{18} Hz, (iv) 1.0×10^8 Hz, (v) 68 kHz. (b) 17 m and 2.3 cm, respectively. $\lambda_{440} = 0.77$ m, a typical size of a musical instrument.

5.7 (a) We denote the fundamental quantities mass, length and time by M, L and T, respectively. Since the dimensions of v are LT^{-1} we have,

$$[\mathrm{LT}^{-1}] \equiv \mathrm{M}^{\alpha}\mathrm{L}^{\beta}[\mathrm{MLT}^{-2}]^{\gamma}.$$

Equating indices of M, L and T we obtain

$$\alpha+\gamma=0,\ \beta+\gamma=1,\ \gamma=1/2,\ \text{giving } v\propto\sqrt{TL/M}.$$

(b) The string with the largest wave velocity will be the thinnest string.

5.8 (a) (i) Using Equation (5.32) find wave velocity $= 50$ m s^{-1}.
(ii) Describing the wave as for example, $y = A\sin(\omega t - kx)$, the maximum value of $\partial y/\partial t = \omega A = 2\pi\nu A = 2.4$ m s^{-1}.

(b) Wave velocity $v = \sqrt{\dfrac{S}{\sigma}} = \sqrt{\dfrac{2.5/0.75}{0.125/0.75^2}} = 3.9$ m s^{-1}.

5.9 (a) At position y, tension $T(y)$ in rope $= yMg/L$, giving velocity $v(y) = \sqrt{yMg/L\mu}$, where $\mu = M/L$. Hence $v(y) = \sqrt{gy}$.
(b) The time it takes the wave to travel a distance δy at y is

$$\delta t = \frac{\delta y}{v(y)} = \frac{\delta y}{\sqrt{gy}}.$$

Hence time taken to travel from the bottom to the top of the rope is

$$\int_0^L \frac{\mathrm{d}y}{\sqrt{gy}} = 2\sqrt{\frac{L}{g}},$$

and time for the return trip $= 4\sqrt{\dfrac{L}{g}}$ which is 2.0 s.

5.10 (a) Using Equations (5.44) and (5.32),

$$P = \frac{1}{2}\mu\omega^2 A^2\sqrt{T/\mu} \text{ which gives } P = 60 \text{ W}.$$

(b) (i) If frequency is doubled, power must increase by a factor of 4 to 240 W.
(ii) If amplitude is halved, power decreases by a factor of 4 to 15 W.

5.11 (a) $I_2 = I_1(r_1/r_2)^2$, since intensity $\propto 1/r^2$.
(b) 5.0 m.

5.12 Total surface area of sphere of radius 1.5×10^{11} m $= 4\pi(1.5\times10^{11})^2$ m^2.
Hence, solar power per square metre on Earth $\approx \dfrac{4\times10^{26}}{4\pi(1.5\times10^{11})^2} \approx 1.4$ kW.
Solar power per square metre on Jupiter $\approx 1.4/5^2$ kW ≈ 56 W.

5.13 (a) From Equations (5.32) and (5.6) obtain $\lambda_2/\lambda_1 = \sqrt{\mu_1/\mu_2}$, giving $\lambda_2 =$ 12.5 cm.
Using Equation (5.54), find $A_2 = 2.0$ cm.
(b) Using Equation (5.55), find $B_1/A_1 = -1/3$, and hence the fraction of power reflected at the boundary $= 1/9$.

5.14 (a) Using Equation (5.52) and the relationship $k \propto n$, obtain ratio of reflected to incident amplitudes $= \dfrac{B_1}{A_1} = \dfrac{n_1 - n_2}{n_1 + n_2}$.

Hence, the fraction of intensity reflected $= \left(\dfrac{n_1 - n_2}{n_1 + n_2}\right)^2 = 0.04$.

(b) Require a thickness of $\lambda/4$, where λ is the wavelength of the light in the MgF_2 coating. $\lambda_{MgF_2} = \lambda_{air}\dfrac{n_{air}}{n_{MgF_2}} = 550\dfrac{1.0}{1.39} = 396$ nm.
Hence, required thickness $= 99$ nm.

(c) For maximum reflection the wave reflected at the glass surface should reinforce the wave reflected at the MgF_2 surface which occurs when the thickness of the MgF_2 coating is $\lambda_{MgF_2}/2 = 198$ nm.

5.15 (a) Equation of motion of central mass, is

$$m\frac{\partial^2 y_r}{\partial t^2} = -T\sin\theta_1 - T\sin\theta_2 \simeq -T\frac{(y_r - y_{r-1})}{a} - T\frac{(y_r - y_{r+1})}{a}.$$

Hence, $\dfrac{\partial^2 y_r}{\partial t^2} = \dfrac{T}{m}\left[\dfrac{(y_{r+1} - y_r)}{a} - \dfrac{(y_r - y_{r-1})}{a}\right]$.

(b) $\dfrac{\partial^2 y}{\partial t^2} = \dfrac{T}{m}\left[\dfrac{y(x+\delta x) - y(x)}{\delta x} - \dfrac{y(x) - y(x-\delta x)}{\delta x}\right]$.

Applying Taylor expansions to the right-hand side of the equation gives,

$$\frac{\partial^2 y}{\partial t^2} = \frac{T}{m}\left[\frac{\delta x\dfrac{\partial y}{\partial x} + \dfrac{1}{2}(\delta x)^2\dfrac{\partial^2 y}{\partial x^2}}{\delta x} - \frac{\delta x\dfrac{\partial y}{\partial x} - \dfrac{1}{2}(\delta x)^2\dfrac{\partial^2 y}{\partial x^2}}{\delta x}\right] = \frac{T}{m}\delta x\frac{\partial^2 y}{\partial x^2}$$

Hence $\dfrac{\partial^2 y}{\partial t^2} = \dfrac{T}{\mu}\dfrac{\partial^2 y}{\partial x^2}$, where $\mu = m/\delta x$.
(Note: As $\delta x \to 0$, $\delta m \to 0$ so that μ remains finite.)

SOLUTIONS 6

6.1 (a) $v = 44.3$ m s^{-1}, $\lambda = 1.0$ m and $\nu = 44.3$ Hz.

(b) Maximum value of $\left(\dfrac{\partial y}{\partial t}\right) = A\omega = 8.35$ m s^{-1}.

Maximum value of $\left(\dfrac{\partial^2 y}{\partial t^2}\right) = A\omega^2 = 2.32 \times 10^3$ m s^{-2}.

6.2 (a) $\lambda = 0.27$ m.
(b) $L = \lambda/2 = 0.135$ m.
(c) Same frequency, 262 Hz, but $\lambda_{air} = 1.3$ m, since velocities in wire and air are different.

6.3 (a) $(n+1)\lambda_{n+1} = n\lambda_n = 2L$.
Hence, $n = 4$, $n + 1 = 5$, $L = 1.1$ m.

(b) Separation of cold spots $= \lambda/2$, giving $\nu = 3 \times 10^{10}$ Hz for $c = 3 \times 10^8$ m s^{-1}.

6.4 Make use of $\cos(\alpha + \beta) + \cos(\alpha - \beta) = 2\cos\alpha\cos\beta$.
Minimum amplitude $= (1 - R)A$ at node of $2RA\cos\omega t\cos kx$.
Maximum amplitude $= (1 - R)A + 2AR$ at antinode of $2RA\cos\omega t\cos kx$.

6.5 (a) $\nu_2 = 880$ Hz, $\nu_3 = 1320$ Hz; velocity remains the same.
(b) Number of harmonics $n \times 440 < 15\ 000$, giving $n = 34$.
(c) From Equations (5.6) and (6.12) obtain, $L_2 = L_1\nu_1/\nu_2 = 26.9$ cm and so string should be fingered at 5.1 cm from the end of the string.

6.6 (a) One octave corresponds to a factor of 2 increase in frequency and n octaves correspond to a factor of 2^n. For frequency range ν_1 to ν_2, $\nu_2/\nu_1 = 2^n$.
Hence, $n \leq \dfrac{1}{\log 2}\log\left(\dfrac{\nu_2}{\nu_1}\right)$, giving $n = 9$ complete octaves for $\nu_1 = 20$ Hz and $\nu_2 = 15$ kHz.
(b) $\nu_1 = 2 \times 10^5$ Hz, $\nu_2 = E/h = 2.4 \times 10^{20}$ Hz, giving 50 octaves between them.

6.7 From Equations (5.6) and (6.12) we have $\nu = \dfrac{1}{2L}\sqrt{\dfrac{T}{\mu}}$.

$$\frac{\delta\nu}{\delta T} \simeq \frac{d\nu}{dT} = \frac{\nu}{2T}, \text{ giving } \frac{\delta\nu}{\nu} = \frac{1}{2}\frac{\delta T}{T}.$$

6.8 (a) From $\nu = \dfrac{1}{2L}\sqrt{\dfrac{T}{\mu}}$ and $\mu = \pi\rho(d/2)^2$, where ρ is density of material, obtain $d_2 = d_1\nu_1/\nu_2$, giving $d_2 = 1.2$ mm.
(b) Using Equation (6.18) obtain total force on neck ≈ 600 N for six strings.
(c) From above, $d_2 = d_1\sqrt{\rho_1/\rho_2}$, giving a diameter of 0.73 mm for nylon string.

6.9 (a) $\dfrac{M}{3}\dfrac{d^2y_1}{dt^2} = -T\sin\theta_1 + T\sin\theta_2 = -T\dfrac{y_1}{L/4} + T\dfrac{(y_2 - y_1)}{L/4}$, etc.
Assuming $y_1 = A\cos\omega t$, $y_2 = B\cos\omega t$ and $y_3 = C\cos\omega t$, obtain

$$A(\omega^2 - 2\alpha) + B\alpha = 0,$$
$$A\alpha + (\omega^2 - 2\alpha)B + C\alpha = 0,$$
$$B\alpha + C(\omega^2 - 2\alpha) = 0,$$

where $\alpha = 12T/LM$. Hence,

$$\begin{bmatrix} (\omega^2 - 2\alpha), & \alpha, & 0 \\ \alpha, & (\omega^2 - 2\alpha), & \alpha \\ 0, & \alpha, & (\omega^2 - 2\alpha) \end{bmatrix}\begin{bmatrix} A \\ B \\ C \end{bmatrix} = 0$$

For non-zero solutions we require the determinant to vanish, giving

$$(\omega^2 - 2\alpha)(\omega^4 - 4\alpha\omega^2 + 2\alpha^2) = 0.$$

Hence, $\omega_{1,3}^2 = \dfrac{4\alpha \pm \sqrt{8\alpha^2}}{2} = (2 \pm \sqrt{2})\alpha,\ \omega_2 = 2\alpha.$
This gives the frequencies,

$$\nu_1 = 0.42\sqrt{T/LM},\ \nu_2 = 0.78\sqrt{T/LM} \text{ and } \nu_3 = 1.02\sqrt{T/LM}.$$

(b) For a string we obtain from Equations (5.6) and (6.12), $\nu_n = \dfrac{n}{2}\sqrt{\dfrac{T}{LM}}$, giving $\nu_1 = 0.5\sqrt{T/LM}$, $\nu_2 = \sqrt{T/LM}$ and $\nu_3 = 1.5\sqrt{T/LM}$, which can be compared with the normal frequencies of the three-mass system.

6.10 (a) $\Delta\nu = c/2L = 1.5 \times 10^8$ Hz.
Therefore number of modes $= 4.5 \times 10^9/1.5 \times 10^8 = 30$.
(b) $L = 3.3$ cm for just one mode to exist.

6.11 Modes that will not be excited are those with a node at one-third the length of the string, e.g. $n = 3$, 6 and 9.

6.12 From Equation (6.37) we have

$$A_n = \frac{2}{L}\int_0^L \mathrm{d}x\, \alpha x \sin\left(\frac{n\pi}{L}x\right),\quad n = 1, 2, \ldots.$$

Using standard integral $\int \mathrm{d}x\, x \sin ax = \dfrac{1}{a^2}\sin ax - \dfrac{x}{a}\cos ax$, find

$$A_n = \frac{2\alpha}{L}\left[\left(\frac{L}{n\pi}\right)^2 \sin\left(\frac{n\pi x}{L}\right) - \left(\frac{xL}{n\pi}\right)\cos\left(\frac{n\pi x}{L}\right)\right]_0^L = -\frac{2\alpha}{Ln\pi}\cos n\pi.$$

Hence, $A_1 = 2\alpha L/\pi$, $A_2 = -2\alpha L/2\pi$ and $A_3 = 2\alpha L/3\pi$, giving

$$f(x) = \frac{2\alpha L}{\pi}\left[\sin\left(\frac{n\pi}{L}\right) - \frac{1}{2}\sin\left(\frac{2n\pi}{L}\right) + \frac{1}{3}\sin\left(\frac{3n\pi}{L}\right) - \cdots\right].$$

6.13 (a) If string is displaced a distance x, the force acting at the mid point is

$$-2T\sin\theta \simeq -4Tx/L.$$

Work done in moving the mid point a further distance $\mathrm{d}x$ is $4Tx\mathrm{d}x/L$.
Hence total work done for displacement d is $\dfrac{4T}{L}\displaystyle\int_0^d x\mathrm{d}x = \dfrac{2T}{L}d^2$.

(b) Using Equations (6.27), (6.8) and (5.32) obtain $E_n = \dfrac{1}{4}\dfrac{n^2\pi^2 T A_n^2}{L}$.

The lowest three excited modes are $n = 1$, 3 and 5 with amplitudes of $8d/\pi^2$, $-8d/(3\pi^2)$ and $8d/(5\pi^2)$, respectively. Hence, the sum of the energies of these three modes is

$$\frac{T(8d)^2}{4L\pi^2}\left(1 + \frac{1}{3^2} + \frac{1}{5^2} + \ldots\right) = 1.87\frac{Td^2}{L}.$$

Hence, the fraction of the total energy $= 1.87/2 = 93.5\%$.

6.14 The function is often described as a *square wave* function. The more terms that are included in the series, the better the approximation to a square wave.

6.15 (a) Using de Broglie, $\lambda_n = 2L/n = h/p_n = h/\sqrt{2mE_n}$, giving $E_n = \dfrac{n^2h^2}{8mL^2}$.
(b) Putting $n = 1$, $E_n = 1.5 \times 10^{-18}$ J ≈ 10 eV.

SOLUTIONS 7

7.1 (a) Since 10 bright fringes span 1.8 cm, fringe separation = 0.20 cm and hence using Equation (7.15), $\lambda = 600$ nm.
(b) (i) Using Equation (7.13), the distance between the two nth bright fringes $= 2n\lambda L/a$ which is equal to 7.6 mm for $n = 2$. (ii) Similarly, the distance between the two $n = 2$ dark fringes is 9.5 mm.

7.2 Angular separation of fringes $\theta = \lambda/a$.
Wavelength of light in medium with refractive index n is given by

$$\lambda_{\text{medium}} = \frac{\lambda_{\text{air}}}{n}.$$

Hence, $\dfrac{\theta_{\text{medium}}}{\theta_{\text{air}}} = \dfrac{\lambda_{\text{medium}}}{\lambda_{\text{air}}} = \dfrac{1}{n}$, giving $\theta_{\text{medium}} = 0.03°$.

7.3 Before the film is inserted, the $n = 15$ bright fringe occurs at distance $d = 15\lambda L/a$ from the central ($n = 0$) bright fringe. After the film is inserted, the *optical path* of wavelets from the covered slit is increased by an amount equal to $(n-1)t$, where n is the refractive index of the film and t is its thickness. At the new position of the central fringe, the amount $(n-1)t$ must be compensated by the distance $a\sin\theta \simeq ad/L$, cf. Figure 7.4. Hence $t = \dfrac{ad}{(n-1)L}$ with the value 1.25×10^{-5} m.

7.4 (a) Angular divergence of sunlight on Earth $\approx$ angle subtended by Sun at the Earth $\approx 1.4 \times 10^6/1.5 \times 10^8 \approx 1 \times 10^{-2}$ rad.
(b) From Equation (7.20), divergence of light from source $\ll 2\lambda/a \ll 1.5 \times 10^{-3}$ rad. This value is much smaller than the divergence of sunlight.

7.5 First minimum of diffraction pattern from slit of width d occurs at angle θ given by $\sin\theta = \lambda/d$. The nth bright fringe from the two-slit interference pattern occurs at angle θ given by $\sin\theta = n\lambda/a$, where a is slit separation. Hence, $\dfrac{\lambda}{d} = \dfrac{15\lambda}{a}$, which gives $a = 0.90$ mm.

7.6 Constructive interference occurs when path difference $= n\lambda$, cf. Equation (7.1). For this problem, use of Equation (7.10) leads to constructive interference when $\sin\theta = n\lambda/a$, where a is the separation of the loudspeakers, cf. Figure 7.4. The full circle corresponds to the angle θ going from 0 to 360°. In the first quadrant $\theta = 0°$ to 90°. Since $\sin\theta \leq 1$, $n\lambda/a \leq 1$ giving $n = 0$, 1, 2, 3 or 4 since $a/\lambda = 4$, with $\lambda = 0.34$ m and $a = 1.36$ m. It follows that there will be a total of 16 maxima around the complete circle.

7.7 (a) Distance between successive maxima $= \lambda/2 = (1.0 \times 10^{-3}/4000)$ m, giving $\lambda = 500$ nm.

(b) For given conditions, $m\dfrac{\lambda_1}{2} = (m+1)\dfrac{\lambda_2}{2} = x$, where x is movement of mirror and m takes integer values. This leads to $x = \dfrac{\lambda_1\lambda_2}{2(\lambda_1 - \lambda_2)}$, which has the value of 0.29 mm for given values of λ_1 and λ_2.

7.8 The gas, of refractive index n, in the gas cell increases the *optical path length* in one arm of the interferometer by an amount equivalent to a mirror movement of $t(n-1)$, where t is the length of the gas cell.

$$\therefore t(n+1) = m\frac{\lambda}{2}, \quad \text{where } m = 90.$$

This gives $n = 1.00036$ for given values of t and λ.

7.9 Sound waves are reflected off successive steps of the amphitheatre and they interfere constructively when $n\lambda = 2L$, where L is the length of the step and $n = 1, 2, 3, \ldots$. This is the same expression as for standing waves on a stretched string.

$$\nu_n = \frac{nv}{2L} = 340 \text{ Hz, } 680 \text{ Hz, } 1020 \text{ Hz, etc., for a value of } L = 0.5 \text{ m.}$$

7.10 (a) From Equation (7.35) find $L \approx 4$ km for typical values of the parameters involved. There would be no interference effects since the light sources are independent and not coherent with each other.

(b) 2.8×10^{-7} rad or $1.6 \times 10^{-5\circ}$.

7.11

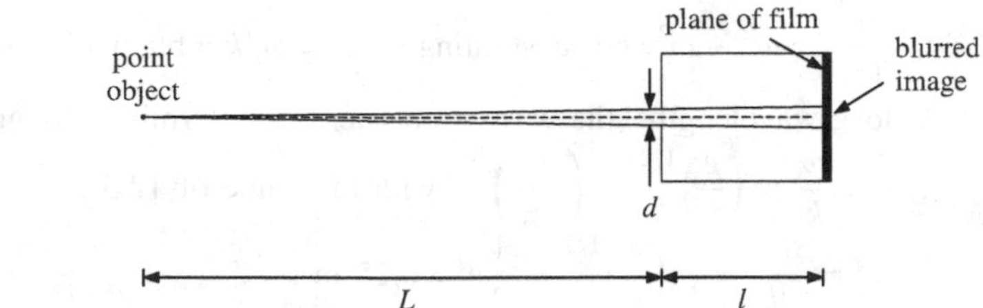

(a) A point source should produce a point image. However, if the pin hole has a finite diameter d, rays of light from the source will produce a blurred image of finite extent as illustrated by the figure. When $L \gg l$, the diameter of this blurred image $\approx d$ and the larger the value of d the greater the amount of blurring. However, the pin hole will produce a diffraction pattern at the film plane whose width $\propto 1/d$, cf. Equation (7.35), i.e. the smaller the value of d the greater the amount of blurring due to diffraction at the pin hole.

(b) For the total amount of blurring to be minimised, the two effects should each produce about the same amount of blurring. This means: diameter of pin hole $\approx$ width of diffraction pattern.

This gives $d \approx \dfrac{2.44\lambda l}{d}$ and hence, $d \approx \sqrt{2.44\lambda l}$.

(c) $d \approx 0.45$ mm.

SOLUTIONS 8

8.1 Beat frequency $\Delta\nu = \nu_2 - \nu_1 = c\left(\frac{1}{\lambda_2} - \frac{1}{\lambda_1}\right) = c\left(\frac{\lambda_1 - \lambda_2}{\lambda_1\lambda_2}\right) \simeq \frac{c\Delta\lambda}{\lambda_1^2}$ to a very good approximation, giving $\Delta\lambda = \lambda_1^2\Delta\nu/c$.
$\therefore \lambda_2 = \lambda_1 - \Delta\lambda = (766.49110 - 0.00090)$ nm, for given values of λ_1 and $\Delta\nu$.

8.2 (a) The second harmonic of the E string has the same frequency (1320 Hz) as the third harmonic of the A string. If one string is slightly out of tune, beats are produced. (b) A beat frequency of 4 Hz means the frequency of the second harmonic of the (sharp) E string = 1324 Hz and hence, the fundamental frequency = 662 Hz.

8.3 A wave pulse travels at the group velocity. The frequencies of the modes of vibration of a taut string depend on the phase velocity, cf. Equation (6.8).

8.4 (a) (i) $\frac{\omega}{k} = \frac{\alpha}{\lambda} = \frac{\alpha k}{2\pi}$, giving $\omega = \frac{\alpha k^2}{2\pi}$, where α is a constant.

Hence, $\frac{d\omega}{dk} = \frac{2\alpha k}{2\pi} = 2v$.

(ii) $\frac{\omega}{k} = \frac{\alpha}{\lambda^{1/2}} = \frac{\alpha k^{1/2}}{(2\pi)^{1/2}}$, giving $\omega = \frac{\alpha k^{3/2}}{(2\pi)^{1/2}}$, where α is a constant.

Hence, $\frac{d\omega}{dk} = \frac{3}{2}v$.

(b) For $\omega = ck$, $\frac{\omega}{k} = c$ and $\frac{d\omega}{dk} = c$.

(c) $\frac{c^2}{v^2} = 1 - \frac{\omega_o^2}{\omega^2}$.

$\therefore \frac{\omega^2 c^2}{v^2} = \omega^2 - \omega_o^2$ and substituting for $v = \omega/k$, obtain $\omega^2 = \omega_o^2 + c^2k^2$.

8.5 (a) At long wavelengths, the wave properties are determined by gravity.

(i) $v = \frac{\omega}{k} = \left(\frac{g}{k}\right)^{1/2} = \left(\frac{g\lambda}{2\pi}\right)^{1/2}$ with the value of 12.5 m s^{-1},

(ii) $v_g = \frac{d\omega}{dk} = \frac{1}{2}\left(\frac{g}{k}\right)^{1/2} = \frac{1}{2}v = 6.25$ m s^{-1}.

(b) Starting with Equation (8.34), obtain $v = \left(\frac{g\lambda}{2\pi} + \frac{2\pi S}{\lambda\rho}\right)^{1/2}$.

Then $\frac{dv}{d\lambda} = \frac{1}{2v}\left(\frac{g}{2\pi} - \frac{2\pi S}{\lambda^2\rho}\right)$, and minimum phase velocity occurs at $\lambda = 2\pi\left(\frac{S}{\rho g}\right)^{1/2}$ which equals 1.7×10^{-2} m, for given values. Using this value of λ, the minimum value of v is 0.23 m s^{-1}.

8.6 (a) Spacing of antinodes = $\lambda/2$ giving $\lambda = 0.5$ mm.
Using Equation (5.6), obtain $v = 0.675$ m s^{-1}.

Since wave properties are determined by surface tension, $v = \left(\frac{Sk}{\rho}\right)^{1/2}$,

giving $S = \frac{v^2\rho\lambda}{2\pi} = 0.49$ N m^{-1}.

(b) In the limit of short wavelength, $v_g = \frac{3}{2}v = 1.0$ m s^{-1}.

8.7 (a) $n = \dfrac{c}{v} = A + \dfrac{B}{\lambda^2}$, giving $\dfrac{c}{\omega} = \dfrac{A}{k} + \dfrac{Bk}{(2\pi)^2}$ from which we obtain

$$\frac{\mathrm{d}\omega}{\mathrm{d}k} = \frac{\omega^2}{ck^2}\left(A - \frac{B}{\lambda^2}\right) = \frac{v}{n}\left(A - \frac{B}{\lambda^2}\right).$$

$$\therefore \frac{v_g}{v} = \frac{(A - B/\lambda^2)}{(A + B/\lambda^2)}.$$

(b) For given values, $v_g/v = 0.87$.

8.8 (a) $y = A\cos(\omega t - kx)$.

$$\frac{\partial^2 y}{\partial t^2} = -\omega^2 A\cos(\omega t - kx); \quad \frac{\partial^2 y}{\partial x^2} = -k^2 A\cos(\omega t - kx).$$

Substitution into $\dfrac{\partial^2 y}{\partial t^2} = \dfrac{T}{\mu}\left(\dfrac{\partial^2 y}{\partial x^2}\right) - \alpha y$ gives $\omega^2 = \dfrac{T}{\mu}k^2 + \alpha$.

(b) The lowest angular frequency is when $k = 0$, giving $\omega = \sqrt{\alpha}$.

(c) From $\omega^2 = \dfrac{T}{\mu}k^2 + \alpha$, find $\dfrac{\mathrm{d}\omega}{\mathrm{d}k} = \dfrac{T}{\mu}\dfrac{k}{\omega}$ and hence $v_g v = T/\mu$.

8.9 From bandwidth theorem, Equation (8.50), obtain

$$\Delta\nu = \frac{\Delta\omega}{2\pi} \approx \frac{1}{\Delta t} \approx 20 \text{ MHz}.$$

8.10 From Equation (5.6) obtain, $\dfrac{\Delta\lambda}{\Delta\nu} \simeq \dfrac{\mathrm{d}\lambda}{\mathrm{d}\nu} = -\dfrac{c}{\nu^2}$, leading to $\Delta\lambda \simeq \left|\dfrac{\lambda^2}{c}\Delta\nu\right|$ and with $\Delta\nu \approx \dfrac{1}{\Delta t}$ from bandwidth theorem, obtain $\Delta\lambda \approx 8$ nm.

8.11 (a) Taking $n = 2$ and using the trignometric relation $\sin 2\beta = 2\sin\beta\cos\beta$, obtain, $A(x, t) = 2a\cos[(x\delta k - t\delta\omega)/2]$.

(b)

$$A(x, t) = a\frac{\sin[n\delta k(x - \alpha t)/2]}{\sin[\delta k(x - \alpha t)/2]}.$$

Using the approximation $\Delta k = (n - 1)\delta k \simeq n\delta k$ when n is very large:

$$A(x, t) = a\frac{\sin[\Delta k(x - \alpha t)/2]}{\sin[\Delta k(x - \alpha t)/2n]} = na\frac{\sin[\Delta k(x - \alpha t)/2]}{\Delta k(x - \alpha t)/2},$$

using the small-angle approximation with n very large.

Index

References to figures are given in italic type; those to tables are given in bold type.

Vibrations and Waves George C. King

Printed in the United States
By Bookmasters